The Restoration of the Nativity Church in Bethlehem

The Restoration of the Nativity Church in Bethlehem

Edited by Claudio Alessandri

CRC Press
Taylor & Francis Group
Boca Raton London New York

CRC Press is an imprint of the
Taylor & Francis Group, an **informa** business

A BALKEMA BOOK

CRC Press/Balkema is an imprint of the Taylor & Francis Group, an informa business

© 2020 Taylor & Francis Group, London, UK

Typeset by Apex CoVantage, LLC

Library of Congress Cataloging-in-Publication Data
Applied for

Published by: CRC Press/Balkema
        Schipholweg 107C, 2316 XC Leiden, The Netherlands

First issued in paperback 2023

ISBN: 978-1-03-257036-5 (pbk)
ISBN: 978-1-138-48899-1 (hbk)
ISBN: 978-1-351-02326-9 (ebk)

DOI: https://doi.org/10.1201/9781351023269

Publisher's Note
The publisher has gone to great lengths to ensure the quality of this reprint but points out that some imperfections in the original copies may be apparent.

Contents

Authors and their affiliations

CLAUDIO ALESSANDRI, Professor of Mechanics of Solids and Structures, University of Ferrara, Department of Engineering, Via Saragat 1, 44122 Ferrara (IT).

MICHELE BACCI, Professor of Medieval Art, University of Fribourg, Switzerland.

MAURO BERNABEI, Researcher, CNR – IBE (National Research Council of Italy – Institute for the Bioeconomy), Via Biasi 75, 38010 San Michele all'Adige (IT).

JARNO BONTADI, Researcher, CNR – IBE (National Research Council of Italy – Institute for the Bioeconomy), Via Biasi 75, 38010 San Michele all'Adige (IT).

GIOVANNI CARATELLI, Researcher, CNR-ISPC (National Research Council – Institute of Science for Cultural Heritage), Rome (IT).

C.D.G. – Community Development Group, Bethlehem, Palestine.

ELISABETTA CONCINA, Art Historian and Restorer, PhD Accademia di Belle Arti, Bologna (IT).

GIUSEPPE ALESSANDRO FICHERA, PhD in Medieval Archaeology, Piacenti SpA.

CECILIA GIORGI, Researcher, CNR-ISPC (National Research Council – Institute of Science for Cultural Heritage), Rome (IT).

LORENZO LAZZARINI, LAMA Laboratory, Iuav University, Venice (IT).

NICOLA MACCHIONI, Researcher, CNR – IBE (National Research Council of Italy – Institute for the Bioeconomy), Via Madonna del Piano 10, 50019 Sesto Fiorentino (IT).

VINCENZO MALLARDO, Associate Professor Mechanics of Solids & Structures, University of Ferrara, Department of Architecture, Via Della Ghiara 36, 44121 Ferrara (IT).

MASSIMO MANNUCCI, Timber structures inspection technical manager, LegnoDOC s.r.l., Via di Colle Ramole 11–50023 Località Bottai – Impruneta (FI), Italy.

MAURIZIO MARTINELLI, Chief Engineer and Designer Legnopiù s.r.l., Viale Borgo Valsugana 11, 59100 Prato (IT).

SABRINA PALANTI, Researcher, CNR – IBE (National Research Council of Italy – Institute for the Bioeconomy), Via Madonna del Piano 10, 50019 Sesto Fiorentino (IT).

PIACENTI S.p.A., Prato (IT).

BENEDETTO PIZZO, Researcher, CNR – IBE (National Research Council of Italy – Institute for the Bioeconomy), Via Madonna del Piano 10, 50019 Sesto Fiorentino (IT).

PRESIDENTIAL COMMITTEE FOR THE RESTORATION OF THE CHURCH OF THE NATIVITY (Palestine).

PAOLA SANTOPADRE, Istituto Superiore per la Conservazione ed il Restauro, Rome (IT).

NICOLA SANTOPUOLI, Associate Professor of Architectural Restoration, La Sapienza University, Rome (IT).

SUSANNA SARMATI, Coservator-restorer, Via del gelsomino 80, Rome (IT).

MARCO VERITÀ', LAMA Laboratory, Iuav University, Venice (IT).

Technical staff and acknowledgements

The high quality of the results achieved is undoubtedly due to the experience, professionalism and dedication of all those who for years worked in the restoration site side by side, overcoming all kinds of difficulties due to the different language, the different geographical provenance, the different cultural and technical training. The restoration site of the church was a magnificent opportunity to bring so many different people together, to make them talk about issues of common interest and to make them work for the achievement of common goals. To all these people, to those mentioned here, but also to the many not mentioned but equally decisive for the achievement of all the set goals, the most heartfelt thanks.

Piacenti S.p.A. – main contractor/Italy

Giammarco Piacenti:	President
Marcello Piacenti:	Project Manager
Paolo Guazzini:	Surveyor
Giorgia Zurla:	Engineer
Silvia Starinieri:	Diagnostic Technician
Christian Piacenti:	Specialist in Structural Reinforcements – Technical Manager
Matteo Piacenti:	Technical Manager Representative
Alessandro Giuseppe Fichera:	Archaeologist
Elisa Frosini:	Archaeologist

Skilled Restorers involved in the worksite: more than 100 experts as Piacenti's employees.

Al Maher – subcontractor/Palestine

Majed Taweel:	General Manager

Community development group "CDG" – construction management and supervision/Palestine

Jamal Araj:	CDG's CEO, Advisory Role
Hussam Salsa':	Project Director
Afif Tweme:	Project Manager
Ibrahim Abed Rabbo:	Resident Engineer
Issa Mourrah:	Senior Structural Engineer
Bandak Bandak:	Administrative Support and Logistics

Chapter 1

Introduction

Presidential Committee for the Restoration of the Church of the Nativity and C. Alessandri

A lot has been said and written so far about the history of the church of the Nativity. Many are the reports, historical documents, scientific articles that can be referenced in retracing the evolution of the church from its origins until today. It is not the intention of the authors to re-propose what has been said so far by authoritative sources and with the utmost abundance of details. The purpose of this book is only to accompany the reader along a path that began in distant 2009 and is now approaching a conclusion, or at least what has been always considered a fundamental goal.

The spiritual significance and the historical and artistic importance of the church of the Nativity have been known since its origin when Emperor Constantine erected it, like other Roman basilicas, as a symbol of Christianity and of the new alliance between the Roman Empire and the Church. Since then the church of the Nativity has always been touched, more or less directly, by the social and political events that have occurred not only in the Middle East but also in Christian Europe.

These events determined fortunes and misfortunes, splendor and decadence of the church and found a correspondence in the many changes made over time both in the structural system and in the decorative parts of the church. The basilica that could be still seen in the first decade of this century was a result of a long period of neglect, due to a paralyzing set of regulations, known as *status quo*, established in 1852 and later confirmed by the Treaties of Paris (1856), Berlin (1878), and Versailles (1919). The *status quo* defined the respective rights of the three religious Communities, the Greek Orthodox, the Franciscan Custody of the Holy Land and the Armenian Orthodox Patriarchate, that still have the management of the church. The *status quo* fixed, and still defines, duties and properties of the three Communities in a very precise way so that the permission to make repairs cannot be considered as a way to assert the hegemonic role in the church of one group instead of the other. Nevertheless, both maintenance and restoration remained difficult in such a context. The *status quo* implied that nothing could be done without a unanimous agreement of the three Communities, which was difficult, if not impossible to be achieved because of the mutual suspicions and rivalries. In 1920, the situation got worse when, during the British Mandate, a new Department of Antiquities was set up with the only duty to examine the repair proposals for Palestinian historical buildings. This was without providing any financial assistance to the Communities, which remained the only responsible parties for the conservation of the Holy Sites.

In consequence of this situation, the basilica remained in the shape received in the last significant restoration, dating back to 1842. This was despite the increasing structural problems of the roof and the progressive deterioration of mosaics, plasters and paintings on the columns. It is worth noting that the sumptuous marble revetments of many internal surfaces had been frequently stolen over the centuries, as witnessed by some pilgrims' accounts. For instance, as stated by the Greek pilgrim Arsenios in 1512, some remnants of the marble revetments were preserved only in the bema and the choir, whereas the nave had been completely deprived of them [1].

The only intervention of conservation in the last century was Gustav Kühnel's cleaning campaign carried out on the decorated surfaces of the church in the early 1980s [2]. An extensive survey was carried out also by Architect William Harvey during the British Mandatory with the aim of restoring significant parts of the church. The interventions were never undertaken; nevertheless, the archaeologic analysis performed provided extremely interesting data for a new re-reading of the history of the church.

Today the basilica is coming back to a new life thanks to an ongoing conservation and restoration program, which has finally overcome the long-lasting stalemate of the *status quo*. This program was the result of two concurrent events, the improvement of the relations among the three religious Communities, through a more open ecumenical dialog, and the engagement of the State of Palestine, as a mediating institution, in promoting mutual agreements. In fact, in 2008, a specific Committee was established by a decree of President Mahmoud Abbas with the aim of defining contents and methods of implementation of a detailed conservation and restoration program for the whole church. The Presidential Committee, still active today, is headed by Ziad Albandak and has several permanent members: Marwan Abdelhamid, Varsen Aghabekian, Nazmi Al Jubeh, Khouloud Daibes, Claudette Habesch, Nabil Kassis, Issa Kassissieh and the present Mayor of Bethlehem City, Anton Salman, who replaced in 2017 the former Mayor Vera Baboun. Its representative in relations with the international experts and the technical staff is Imad Nassar.

In August 2009 an international call for tenders was issued for studies on the state of conservation of the church and for the definition of a comprehensive conservation and restoration program. Applications were expected within December 2009. The call was mainly finalized to the restoration of the roof, which was the most deteriorated part of the church and the one in major need for repairs. However, a detailed analysis of the state of conservation of all the other components of the church (e.g. plastering, masonry, columns, paintings, wall and floor mosaics, floor, wooden architraves and external façades) was also required. The requested deliverables were not only the results of the diagnostic campaign but also guidelines and recommendations with annexed explanatory tables and technical drawings, which the future interventions of restoration should have referred to.

In June 2010, the contract for the execution of the work was awarded to an international consortium of experts (later called Consortium). This included engineers, architects, restorers, experts in the analysis of architectural structures and building materials, historians and archaeologists belonging to Italian Universities and research centres. The Consortium, which was coordinated by Prof. Claudio Alessandri of the University of Ferrara – Italy, as scientific director, was led by CFR (Consorzio Ferrara Ricerche – Ferrara – Italy) as the administrative project manager and was composed of SCDS Pro Inc. (laser scanning survey – Canada), LAP&T – LAAUM (historical and archaeological analysis – University of Siena – Italy), Benecon (analysis of masonry structures – II University of Naples – Italy), CNR Ivalsa (analysis of the roof structures – Florence – Italy), UNIFE (structural analysis – University of Ferrara – Italy), SSBAP (analyses of all decorated surfaces – University of Rome "La Sapienza"), and CDG, Community Development Group (local partner – Bethlehem – Palestine).

In September 2010 the agreement was signed off with the selected Consortium in the presence of the former Palestinian Prime Minister, Dr Salam Fayyad, the representatives of the three concerned Churches (Greek Orthodox Patriarchate, Holy Land Custody of the Holy Land and Armenian Patriarchate) and the President of CFR as a legal representative of the Consortium. The studies were commenced on September 23, 2010, and were divided into three stages: stage I (surveys and documentations – on-site investigations), stage II (studies and assessment – data processing), stage III (recommendations – conservation and restoration proposals and technical guidelines). In March 2011 they ended and the results were officially delivered to the Presidential Committee in July 2011 in the form of a final report containing the required technical recommendations and guidelines to which the designated restorers should have complied.

The work methodology, which was proposed in the report, was aimed at maintaining the integrity of the church and assuring the conservation and protection of the site with all its cultural and religious values. These goals had to be achieved by following as much as possible the principles of restoration, stated in the various Charters of Restoration, ICOMOS and UNESCO documents and by choosing and applying the most appropriate restoration techniques with full respect for the importance and uniqueness of the monument.

For this purpose, the outstanding universal value and authenticity of the Nativity church were deeply considered in all work stages, as a heritage to be preserved and transferred to future generations.

In more detail, as shown in the report through examples, guidelines and explanatory tables and drawings, the interventions had to respect the following methodological criteria:

a   *the minimum intervention*: it is appropriate (convenient) to intervene just when it is essential for the conservation of the monument;
b   *the distinguishability*: a difference must be maintained between original work and integration. The latter has to be done without affecting the "reading" of the "historical document";
c   *the potential reversibility*: further interventions should be feasible in the future, in replacement of those already done, whenever new and more advanced technologies allow more appropriate remedies;
d   *the expressive authenticity*: all the new additions must be declared to avoid conflict with the original work;
e   *chemical-physical compatibility* between original and new materials.

Even in reducing the seismic vulnerability of the church, i.e. its propensity to suffer damages in the presence of an earthquake, the authors of the report suggested to dedicate more attention in choosing the techniques and procedures which were able to exploit as much as possible the existing architectural components and/or some features of the church (e.g. the roof structures, some masonry walls, particular construction systems etc.). In this way, considerable seismic improvements could be achieved without making significant changes to the main features of the church.

From 2011 and during 2012 the Presidential Committee reviewed the report through independent technical reviewers (ICCROM – International Centre for the Study of Preservation and Restoration of Cultural Property, ARUP – independent firm of designers, planners, engineers, consultants and technical specialists) and with the technical support of Consolidated Contractors Company (CCC). In parallel the Committee worked on securing the necessary funds to implement the restoration program.

According to the results of the survey campaign, the roof and its wooden structures were the parts of the church in very precarious condition and in a significant need for strong interventions of conservation and restoration which could guarantee the stability of the church and stop the damages due to rainwater infiltration in wall mosaics and plastering.

Therefore, the Presidential Committee decided to consider the restoration of the church's roof and windows as a priority intervention. As a result, in April 2013, it launched an international call for its execution. This first work phase was called thereafter Phase I.

The evaluation jury, appointed by the Presidential Committee, was formed by Arup, CCC, the Consortium and a CDG representative. The contract was awarded to Piacenti S.p.A., an Italian company that had participated in the tender with four other companies and whose proposal had obtained the highest scores in both technical and financial offers. The construction management was awarded to Community Development Group (CDG) – Bethlehem with the external support of the Consortium, which had provided the comprehensive research report on the state of conservation of the church and the restoration measures to be taken.

Meanwhile, the basilica, which had already been placed in 2008 on the Watch List of the 100 most endangered sites by the World Monuments Fund, had become on June 29, 2012, the first Palestinian site to be listed as a World Heritage (WH) Site by the WH Committee. The proposal was approved by a secret vote with the only opposition of the United States and Israel. This was according to the UNESCO spokesperson Sue Williams, and following an emergency candidacy procedure that by-passed the usual 18-month process defined for most sites. The church was also placed on the List of WH sites in danger because of the serious damages which had occurred in the roof.

On August 26, 2013, the contract for the restoration of roof and windows was signed by the Committee's Chairperson Ziad Albandak and the Contractor. This was during a ceremony held in the Presidential Palace of Bethlehem in the presence of Dr Rami Al Hamdallah, the current Palestinian Prime Minister, the representatives of the three religious Communities and other Ministers and officials.

Since the commencement of the works on September 15, 2013, the Committee received generous donations from different parts of the world. This encouraged the Committee to continue the restoration of other parts of the church with reference to a list of priorities previously agreed with the Consortium. These additional works included the restoration of the narthex and the narthex and basilica doors (Phase II), the external stone façades, internal wall plastering, wall mosaics and wooden architraves (Phase III), lighting and smoke detection systems, the restoration of all the columns of the nave along with their painted surfaces (24 out of 50 columns) and part of the floor mosaics (Phase IV). For the entire duration of the works, C. Alessandri and some members of the previous Consortium, such as B. Pizzo, S. Sarmati, V. Mallardo and G.A. Fichera offered their contribution in validating the technical documentation provided by Piacenti Company and in periodically supervising, on behalf of the Presidential Committee, the technical execution of the interventions. Meanwhile the Committee organized a campaign called "Adopt a Column" to attract more donors, especially from small businesses or families, in order to complete the restoration of the remaining columns and paints.

The key stages through which the entire restoration process was developed, from the beginning until today, are listed below. Each work phase will be described in detail in the next chapters:

Phase I:

2013:	organization of the worksite
2013–2015:	restoration of roof and windows

Phase II:

2014–2015:   doors of narthex and basilica
2014–2017:   narthex

Phase III:

2014–2017:   external façades
2014–2016:   plastering
2015–2016:   wall mosaics
2015–2016:   architraves

Phase IV:

2016–present: columns and paints
2017:   lighting and smoke detection system
2018– present: floor mosaics

Future Phases:

Remaining columns and paints
Remaining floor mosaics
Stone floor
Seismic reinforcement of the north and south corners
Seismic reinforcement of the south wall

On July 2, 2019, the World Heritage Committee, meeting in Baku, decided to remove the church of the Nativity and the Pilgrimage Route (Bethlehem) from the List of World Heritage in Danger. The Committee said its decision was due to the high quality of work carried out on the Nativity church, the restoration of its roof, exterior façades, mosaics and doors. It also welcomed the shelving of a project to dig a tunnel under Manger Square and the adoption of a management plan for the conservation of the site.

The restoration of the church of the Nativity has been and will always be an event of extraordinary cultural, religious and political importance. At present the church is the only Byzantine monument in all Palestine handed down in its almost original form, for which a global restoration was designed and above all completed with such a vast use, never seen in precedence, of economic and human resources, diverse and highly qualified professional skills and advanced technologies and instruments.

Since the initial study campaign, and also during the subsequent phases of design and implementation of the interventions, the work was always based on the closest collaboration between the various disciplines and skills that had to be involved. This strong interdisciplinary approach has allowed observers to have a continuous, global perception of the work that was being done, not relegated to the narrow sphere of the single problem, but extended to all the plurality of actions that were taking place or that had to be planned. This has allowed significant advantages, such as a coordination between the various interventions (with consequent saving of time, human resources and money), an optimization and a greater reliability of the choices made or to be made (thanks to a comparative analysis of the various problems involved), and finally a sharing of the information, particularly necessary in a complex and articulated working context such as that of the church of the Nativity.

After the archaeological campaign of Harvey in 1934 [3], the results of which offered the matter for further studies [4][5], the current program of interventions in the church represents the first new opportunity to resume an interrupted cultural discourse. It also offers the whole world new material for a deeper, and above all more complete, knowledge of the history and peculiarities of the church. As a matter of fact, the archaeological excavations carried out by Harvey remained incomplete. Begun in April 1934, they concerned only parts of the nave, the northern part of the transept and part of the choir and ceased very early to allow the Christmas celebrations in the same year. Thus, the findings only provided partial and fragmentary information, which was however still very useful for a more correct knowledge of the past history of the church. In fact, they provided material for new and more detailed investigations and made it possible to give answers to some questions that had remained unresolved until then.

Even the scientific production of the last century reflects in some way this fragmentary nature of knowledge through a variegated series of articles in scientific journals and chapters of books with diversified contents. Moreover, these were available in a multiplicity of languages, which sometimes made it difficult for an easy and quick understanding. Very often they are scientific works that do not always interact with each other, either for linguistic difficulties or for a sort of rivalry between different schools [2]. On the contrary, by summarizing the work experiences carried out in the last decade throughout the church in a coordinated and interactive manner, this book tries to provide, through the description of the restoration interventions made and using the big amount of information collected on-site, a more comprehensive and organic knowledge of the church as a whole. Therefore, in exposing the state of conservation of each part of the church (roof structures and covering, masonry, plasters, mosaics, columns and paintings) and the interventions of restoration carried out, reference will be made to all those archaeological findings discovered on-site during the work phases and through which it will be possible to relocate in their place some tesserae of the large mosaic representing the history of the church. It is hoped that every reader has something to learn from the results of this restoration campaign and that a fruitful intellectual exchange can be established among architects, restorers, art historians and engineers, like the one which took place daily on the scaffolding of the church throughout these years.

The restoration of the church has affected and will deeply affect the relations between the three religious Communities that run the church by further improving the ecumenical dialog which already started many years ago. Every decision regarding types and modalities of intervention was agreed with the representatives of these Communities and the Presidential Committee, without neglecting the peculiarities and specific needs of each religious tradition. However, the restoration interventions carried out and the need to preserve over time what has been restored will impose some changes in certain exterior manifestations of faith of some Communities. Moreover, more importantly, the interventions will require the acceptance of common behavioral norms for the future maintenance of the church, beyond what was established by the current *status quo*. This will undoubtedly contribute to unblocking the stalemate that the latter created around the middle of the 19th century and to make the relations among the three Communities smoother and more oriented to a greater practicality. Encouraging signs have already occurred after the completion of the restoration of the mosaics. Now, the rediscovered beauty of these mosaics requires greater respect and, therefore, the adoption of appropriate behaviors and measures by every part in order to protect their integrity over time. The entire maintenance plan of the church, necessary to preserve the state of conservation and the integrity of the church over time, as even requested by

UNESCO, will be a further opportunity for new and more intense relationships. However, during the restoration campaign, many work phases were already facilitated and accelerated by agreements made instantly among the three Communities on the basis of mutual trust, common sense and above all in the awareness of cooperating to achieve a goal of universal value.

On the occasion of the restoration of the church, various possibilities for a cultural and social development for the local community have emerged, some of which have already found forms of implementation while others are still under study. The church of the Nativity is nonetheless a container of inexhaustible potentialities and as such must be constantly kept into account for all that it can offer in any area. For example, during the first study campaign, some members of the Consortium, by the end of the daily work, gave lectures in the classroom to local professionals and University students on the problems they had faced within the church and on the technical solutions that were being prepared in cooperation with the Presidential Committee. In fact, the gap that still exists with Western Countries, such as Italy, on the issues of conservation and restoration of the historical, artistic and architectural heritage is undeniable, although the experience gained throughout these years in the church of the Nativity has undoubtedly served to grow the interest and sensitivity towards the problems that conservative restoration poses and also to raise among the local communities the awareness of the enormous cultural and artistic heritage they are bound to preserve over time.

In this regard, it is worth underlining the intense work done by some local institutions for more than two decades in documenting the Palestinian heritage and culture by identifying buildings and sites with greater need for restoration and maintenance. With such institutions it is also possible to share the vision of the architectural restoration as a social and economic incubator, to the extent that a restoration serving the public creates jobs and strengthens the community's identity.

In order to disseminate these ideas and foster these feelings, the Consortium formulated an educational program that should have been implemented through theoretical lectures in the classroom and guided visits to the site during the restoration works. The aim of this program was to allow possible users, such as University students, professionals, representatives of local institutions and cultural centres to find a full correspondence between principles and theories of restoration, traditional and innovative restoration techniques and their applications in the real body of the church. In this way it would have been possible to achieve what has always been the dream of every restorer, i.e. transforming the building site into a professional training school where theory and practice can find a mutual integration and enrichment. Unfortunately, due to economic hardships and some logistical and organizational difficulties, the implementation of this program was not possible. However, it could still be implemented even with different modalities and times. To this end, an important role can be played by the communication strategy. This strategy advocates that the universal value of the church, the work done so far, the history, the investigations, the methodologies, the technologies used, the principles and criteria that underlie the choices, the management and maintenance programs and the information on possible social and economic implications can all be shared with all potential stakeholders.

Moreover, the local authorities have committed to take some measures to promote educational programs, spread knowledge and ensure that the church's restoration experience does not end in itself. But rather it can be a continuous source of new and increasingly interesting initiatives to benefit the site and the entire local community. These measures might be in the form of publications in the form of brochures, newsletters with periodic reports on the activities within the site or related to it, information brochures, multilingual websites in addition to or as integration of the existing one (http://nativityrestoration.ps), a link in the web page of the city of Bethlehem, events organized within the church and/or in collaboration with the municipality to increase the awareness of the importance of the site and its conservation in particular, exhibitions based on selected topics concerning the site, the choice of a place in the vicinity of the church where conferences and meetings can be held, networks with other sites and institutions all over the world for mutual exchange of experiences, international partnerships to increase levels of cooperation, the inclusion of the church in more comprehensive touristic tours etc.

These are measures that should also be part of the management and maintenance program. The Presidential Committee and the local authorities will have to prepare such measures for the conservation of the monument and for what has been restored so far, but above all for the inclusion of the church in a network of local and international relations that foster its knowledge and underline its historical, artistic and religious importance and the decisive role that this monument can play in the social and cultural development of local community.

Each chapter of the book refers to a specific part of the church and describes in detail the conservation and restoration interventions carried out on the basis of previous historical studies, surveys, analyses of the state of conservation, tests on materials carried out on-site and in laboratories. The topics are treated with scientific rigor but at the same time in a very discursive way and by using a simple and easily understandable language even by those who for the first time approach the problems of the restoration of historical buildings. In a narrative style, which is mainly based on a reasoned description of the interventions performed, the authors will try to exploit the reader's ability to immediately understand the nature of the problems and the reasons behind the choice of particular techniques or procedures. With reference to some interventions and at the end of to some chapters, some simple guidelines are also provided for the maintenance of what has been restored. They absolutely do not claim to be exhaustive, also because a complete maintenance plan would require further analyses, more in-depth checks on-site, the exact definition of the check procedures and, above all, further agreements with the Communities. More detailed information and specific references in this regard will be contained in the Management and Maintenance Plan for the church of the Nativity, an integrated action plan defining goals and measures for the protection, conservation, use and development of the site that the Presidential Committee will have to provide according to what required by UNESCO to all properties inscribed on the World Heritage List.

The reader will notice that the way of presenting the bibliographic references in the various chapters is not generally the same for all the chapters. Since the book is the collection of contributions from experts in different disciplines, sometimes even very distant from each other, each one with its own rules and traditions, it was preferred to leave the authors free to insert the bibliographic references in the way considered most appropriate to the type of topics covered. Moreover, for the non-purely specialist feature of this book, in the bibliographies of the chapters concerning technical interventions it was preferred to mention mainly, and sometimes exclusively, the technical standards followed. Detailed references to previous scientific works on those matters can be found in the various scientific articles and books published so far on the restoration of the church.

We thank all those who in the last decade have participated in the execution and completion of the restoration of the church, a very important event in the long history of the church and in the interreligious and intercultural relations. Special thanks to the Custody of the Holy Land, the Orthodox Patriarchate and the Armenian Patriarchate, without whose willingness to cooperate in full harmony and unity of purpose these results would never have been obtained. The most heartfelt thanks to President Mahmūd Abbās, who strongly supported this restoration and who has constantly followed every stage of its development. Finally, thanks to all the donors, Governments, public and private institutions and private donors who, with their donations, have made it possible to carry out such a large and demanding work.

References

[1]   Bacci, M., Bianchi, G., Campana, S. & Fichera, G. (2012) Historical and archaeological analysis of the Church of the Nativity. *Journal of Cultural Heritage*, 13, Elsevier, e5–e26.
[2]   Bacci, M. (2017) The Mystic Cave: A History of the Nativity Church in Bethlehem. Rome, Brno, Viella/Masaryk University Press.
[3]   Harvey, W. (1935) Structural Survey of the Church of the Nativity. Bethlehem, Oxford, Oxford University Press.
[4]   Hamilton, R.W. (1947) The Church of the Nativity, Bethlehem: A Guide. Jerusalem, Government of Palestine, Department of Antiquities.
[5]   Bagatti, B. (1952) Gli antichi edifici sacri di Betlemme in seguito agli scavi e restauri praticati dalla Custodia di Terra Santa (1948–1951). Jerusalem, Franciscan Printing Press.

Chapter 2

The restoration site

Piacenti S.p.A., C.D.G. – Community Development Group and C. Alessandri

A restoration site such as the basilica of the Nativity, due to its complexity and the number of operators involved, can be compared in some way to the large work sites of the Gothic-Renaissance cathedrals. In all the stages of the restoration process, from the preliminary analyses until the completion of the interventions, the site was considered extremely delicate and fragile, to be treated with great care and by experienced and specialised teams only. The preservation of the authenticity of the church of the Nativity and the transfer of its universal value to future generations were always taken into account in all the work stages.

2.1 Team configuration

The personnel supposed to work on-site, plan interventions and monitor their execution had to have great technical experience, ability to adapt standard intervention techniques to particular local situations, programming skills and experience in coordinating operators and work phases. In fact, this project has no equal in Palestine in terms of quality and complexity of the interventions, level of technical and artistic knowledge required, specificity of management skills, interdisciplinary character of problems and diversity in provenance (both cultural and geographical) of the operators involved. It therefore represented for the entire work team a great challenge, which was however overcome and won thanks to the close collaboration and deep understanding among members of the Presidential Committee, technical supervisors, restorers and site manager.

The whole team was differentiated according to the different responsibilities that had to be taken:

1   strategic planning and upper management and supervision, played by the Palestinian Presidential Committee (the Client), whose members and representatives were selected according to their professional background and specialization. They were responsible for securing the needed funds for analyses and implementations through organized fundraising campaigns and events, defining the work priorities and the best strategies to get the desired results, overseeing the entire restoration process.

2   ordinary on-site management and supervision, played by a local Palestinian team of Engineers, selected among the most experienced and specialized engineers in this particular field. The team had the responsibility to carry out all management activities and provide all needed technical and managerial assistance to the Client in implementing the approved plans. In particular, they had to check on site the correct implementation of the works approved by the Client on the basis of a list of priorities, check the

correspondence between design shop drawings and their execution, assess the quality of materials, approve safety and site usage plans, guarantee a correct management of time schedules, costs and risks, support the Client in all upper management tasks and in the coordination with the three religious Communities.

3   implementation of the interventions, played by the Italian contractor and his subcontractors, selected carefully among experienced and specialized restorers. The contracting Company appointed a highly qualified project manager, the so-called *capomastro*, with great experience in the restoration field and able to manage the different and multidisciplinary teams of specialized Italian restorers as well as the Palestinian local laborers hired by the local partner. Italian restorers and local laborers lived and worked together for the whole duration of works, overcoming all linguistic and cultural differences in a simple and natural way. This osmosis, predictable but not certain, made it easier to transfer to the local personnel the basic technical know-how and to start *in loco* a training process which will allow in the future local manpower to care of the local historical and artistic heritage with less and less contributions from outside.

4   remote checks and periodic on-site checks, carried out by members of the Italian Consortium who were an essential reference at any stage of the restoration process. In the study stage, they followed up the survey and diagnostic campaign, verified the correctness of the design proposals and their correspondence with the design criteria reported in the main Charters of Restoration and with the Conservation Plan agreed with UNESCO; in the implementation stage they followed up the on-site works through periodic on-site visits or responses to technical reports provided by the site management team. In particular, they provided all needed technical recommendations and guidelines to define restoration methodologies, choose the most appropriate materials and review shop drawings. They also played a major role in all the main meetings held with UNESCO representatives and in all relationships with international institutions.

All parties involved in the implementation of the project were able to match the specific needs imposed by the local situations. Previous knowledge and experiences on different levels were put together within an organized system aimed at assuring the completion of all tasks and the achievement of high-quality results. Managers, designers, researchers, architects, engineers, historians, archaeologists, restorers, laboratory specialists, finance and accounting staff, technicians, skilled laborers and suppliers were all working together as a unique organism, articulated in various and different components but all compatible and closely interacting with one another. Thousands of activities were performed following methodologies and procedures that had been previously controlled at all levels and shared by all the concerned operators.

2.2   Preliminary on-site tests

On October 4, 2013 the scaffolding elements reached the site transported by two 40-ft containers, and they were unloaded into temporary storage areas in front of the Nativity church after getting the approval from the three Churches' representatives. The unloading process, made directly after the arrival of the containers, lasted until midnight to avoid any interference with pilgrims and visitors. Two external storage areas had been prepared with proper fencing and the floor tiles had been protected by using a thick layer of geotextile and pressed

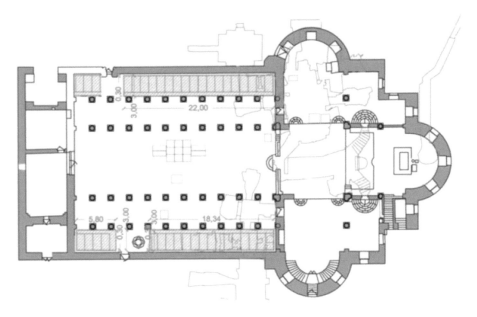

Figure 2.1 Internal storage area (pink colour).

oriented strand boards (OSB). Later, one of the external storage areas was removed, and part of the scaffolding elements were moved inside the church along the outermost aisles (Figure 2.1).

The same protection measures were adopted to protect the church stone tiles from scratches or damage. The internal storage area was also protected by safe fencing. Before building the scaffolding inside the church, a co-ordination was established with the concerned Churches and the Palestinian police to facilitate the contractor's work on a daily basis between 6.00 p.m. and 3.30 a.m. According to the bidding documents, all columns and architraves were protected against possible damages or scratches occurring during the scaffolding assembly. The columns were protected by using a layer of geotextile and vertical wooden boards (Figure 2.2). The same geotextile was used also for the architraves (Figure 2.3).

The wall mosaics at the upper level of the nave (the angels) were firstly covered with a gauze and a non-woven fabric (Figure 2.4), held steady by vertical wooden boards fixed to the walls by means of horizontal wooden fittings (Figure 2.5).

An analogous system of protection was used for the mosaics of the lower level, although less exposed to risks deriving from heavy processing.

Again, these works were done overnight to avoid any disturbance to pilgrims and visitors and to allow the ordinary liturgies during the day. Before such protection works, dilapidation survey was carried out by the contractor in all the work areas to provide high-quality records (photos and videos) of the existing conditions of the site.

The assembling of the scaffolding started in the nave area to cover the entire church. Proper spaces were kept for the safe movements of pilgrims and visitors (Figure 2.6). Particular care was taken to avoid unloading weights on the underground cavities, as will be specified in Section 3.3. Moreover, at the base of each vertical element of the scaffolding, a

Figure 2.2 Preparation for the stone columns' protection.

Figure 2.3 Protection of stone columns and architraves.

sufficiently wide and thick wooden plate was placed in order to spread the load on a wider surface and have, therefore, lower pressures on the floor (Figure 2.6). This scaffolding system, which allowed a high flexibility in assembling, was the first of its type to be used in the region.

Outside the church, several measurements, analyses and tests were made on-site to decide the best location to erect the crane that should have speeded up the restoration process and transported or lifted the materials safely from one place to another. In the first place, the area

Figure 2.4 Application of gauze and non-woven fabric.

Figure 2.5 Final cover with wooden boards.

in front of the main entrance and in the immediate nearby was checked as a possible place for the crane by using electrical resistivity tomography (ERT), a well-known technique for the near-surface exploration of areas characterized by a complex geological setting (Figure 2.7).

As a matter of fact, the presence of antique cisterns in the subsoil of the square in front of the church was already known in the past, as witnessed by the Treatise of Friar Bernardino

Figure 2.6 Scaffolding system inside the church.

Figure 2.7 Electrical resistivity tomography test.

Amico (Bernardino Amico, 1620; Figures 2.8, 2.9) and by photos from the early 1900s (Figure 2.10). It was now confirmed by the ERT and visual inspections (Figure 2.11).

The complexity of the subsoil and the fragility of the stone tiles of the floor oriented the choice towards other solutions. In the end, the results of all the tests carried out showed that the Greek Orthodox garden beside the bell tower was the best location for the crane,

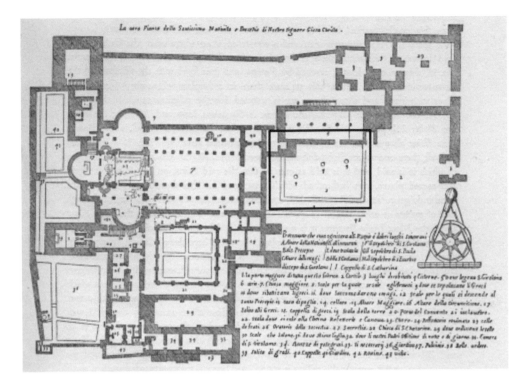

Figure 2.8 Entrance holes to the cisterns. Plan from Bernardino Amico's Treatise.

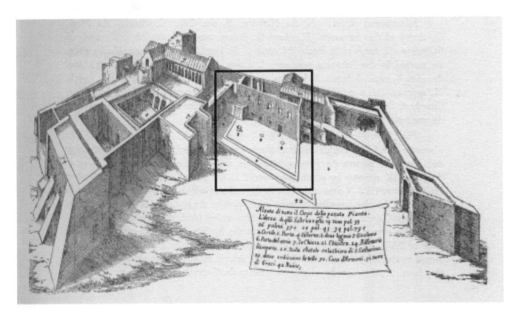

Figure 2.9 Entrance holes to the cisterns. Perspective view from Bernardino Amico's Treatise.

Figure 2.10 Photo of one of the entrance holes to the cisterns.

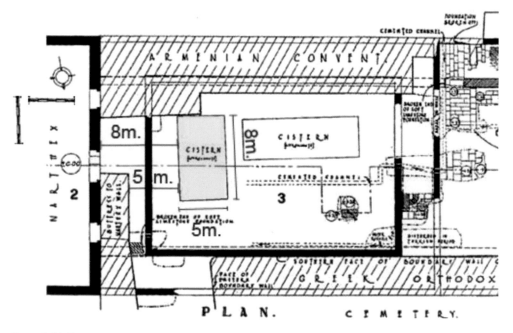

Figure 2.11 Recent survey of the cisterns.

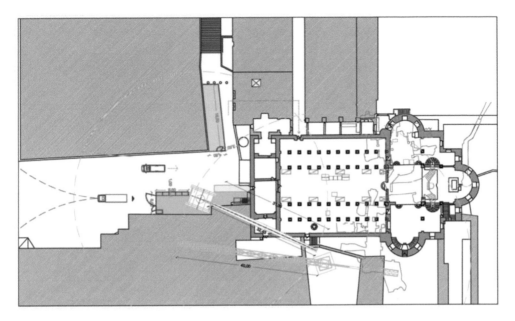

Figure 2.12 General site layout.

which, in this way, was able to serve the maximum area of the roof (Figure 2.12). Further tests including ERT and Laboratory tests on soil samples were performed in the proposed location to assure that no voids or grottos were present and to measure the bearing capacity of the soil where the crane had to be installed.

2.3 Organization of the work site

Before starting with any activity on site, intensive coordination meetings were conducted between all concerned parties to consider the most appropriate procedures for the project implementation and the site organization. Three major issues were deeply discussed and finally overcome with a common agreement:

1 assuring the highest level of safety for the church users and the restoration team,
2 adopting the best working procedures to protect the site from any damage during the works,
3 implementing a work schedule so as to take into account all site requirements, risks and contractual deadlines.

Accordingly, safety plan, work plan and site layout (Figure 2.12) were carefully prepared and approved in order to proceed with the site mobilization.

One of the main requests of the three religious Communities was that the church be kept open during the works and most of the church spaces be safely accessible to everybody. For this reason, special places outside the church were arranged to host noisy or supposed dangerous operations. Moreover, some environmental mitigations (reduction of dust emission,

Figure 2.13 External scaffolding at the western façade.

use of safe chemical materials in the interventions of restoration, limited use of noisy equipment inside, use of stone-textured PVC sheets to hide external storage areas etc.) were implemented by adapting special procedures and techniques to the site requirements in order to avoid any undesired negative impact on the church users or the church itself. All procedures and plans were coordinated with the three concerned Churches respecting the current *status quo*. It is also important to mention that during the whole restoration process all working staff used an external entrance to avoid disturbing visitors and pilgrims. That was possible by means of a scaffolding system at the western façade of the church, which allowed entering the working floors of the internal scaffolding through the terrace of the narthex and the lateral windows of the nave (Figure 2.13).

Before any intervention, all the geometric data of the church were surveyed by using 3D laser scanning, a well-established survey technique, particularly effective when used in historic buildings whose geometries have undergone changes over time, often difficult to survey and evaluate using traditional manual techniques. Furthermore, as is known, 3D laser scanning allows obtaining 3D virtual representations of the actual building, from which it is possible to obtain, at any time and for any part of the building, all the 2D representations (plans and sections) necessary for design stages and successive verifications without any need for further measurements.

2.4   Supply of materials and facilities

The supply of materials, equipment and tools for the site was one of the major challenges faced, as it entailed risks of serious interruptions of the work program in each of the restoration stages. It is worth remembering that most of the materials and equipment needed for the restoration are not available in Palestine, and the restoration sites currently operating, albeit in limited numbers, absorb most of these materials and equipment. As a result, the contractor was often obliged to import materials from Europe, facing dozens of obstacles and logistical restrictions as all imported products must pass through Israeli customs controls. On the other hand, however, this allowed the use of materials having all the certifications required by Italian and European standards.

This major risk was considered in every single operation to be made on site. Any work plan took into account the high possibility to encounter difficulties in the delivering of whatever was needed on site. That means that the order of materials or specific tools/equipment had to be made in advance to secure the availability at the scheduled time of the operation. Such issues burdened the work of the administrative staff and increased, sometimes considerably, the cost of restoration.

Another to the use issue was the peculiarity of the site. Once a container reached the site, the materials were supposed to be moved inside the church. All available scenarios led inevitably to use of one of the church's narrow doors, limited in width and height. Moreover, the visitors' flow had to be kept uninterrupted and safe. Despite these well-known site difficulties, hundreds of tons of material, scaffolding, wood, lime, sand, lead etc. were moved manually inside the church without any accident.

Reference

Bernardino Amico, F. (1620) *Treatise on the Plans and Drawings of the Sacred Edifices of the Holy Land*. apud Pietro Cecconcelli alle Stelle Medicee cum Permissu Superiorum, Florence.

Chapter 3

Roof and windows

3.1 Historical analysis

G. A. Fichera

A curious silence of old sources about the fate of the building envelops the long period between the Empire of Justinian and the Latin conquest of Palestine in 1099. A legend known from a ninth-century Byzantine source emphasises that the basilica had not been destroyed during the Persian invasion of 614, whereas Islamic writers from the 10th and 11th centuries clearly state that, during the Arab conquest of 636, the Caliph Omar had extended his protection over the church of the Nativity. According to some authors, a *mihrab* was built up in the southern apse, and the palm mentioned in the Holy Quran, under which Mary had given birth to Jesus, was preserved in its interior; Islamic pilgrims paid their respects to the place, which was apparently not even damaged during the destructions operated by the Egyptian Caliph al-Hakim in 1009 [1][2]. Legends and historical references find an important validation in the dendrochronological analyses performed during the first diagnostic campaign held in 2009–2010 by the international team (Consortium) led by the University of Ferrara and appointed by the Presidential Committee [3][4]. Such an evidence would seem to indicate that at the arrival of the Crusaders in 1099, the basilica was in fairly good condition. In fact, the enhanced status of the city, the use of the building as coronation church for Balduin I in the year 1100, the elevation of Bethlehem to a bishopric in 1108, the intensification of pilgrimage in the subsequent decades and the further embellishments with new furnishings and ornaments show that despite the various war events that took place over time, the church had managed to remain in a state of preservation adequate to important roles and solemn ceremonies [5]. It should be noted, however, that the same dendrochronological analysis performed on roof samples showed that significant roof restorations had already been carried out at that time. Since wood was a precious material in those areas, difficult to transport, and decisions were made only after a careful and long evaluation, the interventions carried out in those years or in immediately preceding years, confirmed by the dendrochronological analysis, surely testify to a state of roof degradation which, already in that period, was not negligible.

The history of the basilica after the Crusaders is still scarcely investigated [6], the main study being still that by Vincent and Abel [7]. From the 13th to the 19th centuries, the building did not undergo significant alterations: according to both Ayyubid, Mamluk and Ottoman customary law, Christians were allowed to preserve their churches, but they were prevented from both erecting new buildings and embellishing old ones; in order to make repairs, it was necessary to receive a special permission from the sultan himself. This difficult situation caused a lack of maintenance, and the church started falling into a state of decay, as was frequently remarked, from the 14th century onwards, by those same pilgrims who never stopped manifesting their astonishment for the beauty of the church, its paintings, marble incrustations, monumental columns and magnificent roof. The most serious problems concerned the church roof. The medieval one, which was always described as made with several qualities of wood (cedar of Lebanon and cypress) and covered with lead, was by the late 15th century in such a bad state of preservation that rain fell down from its many holes and the pavement was covered with birds' dung, even if, according to some sources, a first restoration had been accomplished in 1435 under the auspices of the Greek Emperor of Trebizond Alexios Komnenos Doukas. Yet the portion of roof overhanging the choir was going to collapse when the Italian pilgrim visited the church in 1474 and saw that the Franciscan friars had been obliged to erect a wooden structure to hold it up [5]. As we are informed by Friar Francesco Suriano, the Franciscan Guardian Giovanni Tomacelli obtained the sultan's permission for the thorough restoration of the roof; this fact is also witnessed by the original firman and Felix Fabri's account. Tomacelli was an Observant friar, and his efforts to restore the basilica's ancient decorum manifested a radical change of attitude, implicitly contrasting that of the previous Conventual administration (as is implied by Suriano's words). He was able enough to obtain sponsorships from the Duke of Bourgogne and the King of England; whereas the latter's money was used for the lead covering, the former's was invested for the making of the new wooden structure. Venetian carpenters and woodcarvers came to Bethlehem to take measurements, and they subsequently made beams out of larch from the Alps. The materials were then transported by ship to Jaffa and thence transferred to Bethlehem by means of camels and oxen; special machines were constructed in order to transport the hugest and longest beams [1].

Sources are silent about the roof in the 16th century, but as early as 1607 and later on in 1617 its condition had become precarious, and the Greek community was allowed to make some substitutions of rotten beams. Yet a much more efficacious intervention took place on the initiative of the Greek Patriarch Dositheos in 1672: thanks to the sponsorship of a rich Greek devotee, Manolakis of Kastoria, it was possible not only to renovate the roof with new beams from Mytilene and a new lead covering but also to make new ornaments in the church. The windows, which had been previously closed with hard stones, were substituted with iron casings and glass; some of the nave walls were plastered, and the entrances to the Nativity grotto were embellished with new marble slabs. Only interventions for the building's ordinary maintenance took place in the 18th century, except for the restoration, in 1775, of a wall, located close to the west entrance, that was going to collapse. In 1834 the basilica was damaged by an earthquake, and already by 1837 the Greek community had received lots of offerings from the devotees to make new embellishments in the narthex. Minor damages were further inflicted with a series of strong aftershocks in 1836 and with the Galilee earthquake of 1837 shortly thereafter. Finally, in 1842, the Sultan Abdul Mecit, answering to the official request of the Greek Patriarch Athanasius III, gave permission to work out a thorough renovation of the wooden roof and its lead coverings. As already mentioned in the Introduction, a set of regulations, known as *status quo*, promulgated by the sultan in 1852 and later confirmed by the Treaties of Paris (1856), Berlin (1878) and Versailles (1919), defined the respective rights of the Greek Orthodox, the Armenians and the Latins of the Franciscan Custody of the Holy Land in the use and ordinary maintenance of the church. After centuries of rivalries among the three religious Communities, during which the official permission given by the authorities to make repairs to the building was regarded by each Community as a way to assert its own *praedominium,* or hegemonic role in the holy site, the compromise found in the 19th century fixed the duties and properties of the three groups in a very precise way. Nevertheless, both maintenance and interventions of restoration continued to raise difficulties in such a context. The *status quo* implied that nothing could start without a unanimous agreement of the three groups, which was made hard, if not impossible, by mutual suspicions and rivalries. The situation was further complicated by the ordinance issued in 1920 by the British Mandate for Palestine that set up a new Department of Antiquities. Whereas the latter had the duty to approve all repairs to historical buildings, it was in no position to provide any financial assistance, and the religious Communities only had the responsibility for the conservation of the holy sites. Problems arose periodically, as e.g. in 1926, when the Latins and Armenians attempted to share the costs for some urgent repairs to the roof but were prevented by the Greeks, who exercised in this way their rights of proprietorship on the upper parts of the building, as established by the *status quo* [6]. The outcome of this situation was that the basilica was frozen in the shape received after the last significant restoration, dating back 1842, and very little was done to solve the structural problems of the roof until September 2009, date of commencement of the first survey and diagnostic campaign by the Consortium. Starting from December 2013 until June 2014 wood experts appointed by the contractor analyzed more extensively and in more detail the conditions of the wooden structure supporting the roof, and also a number of on-site tests were carried out in the nave and the aisles to get information on the geometry of the wooden elements, characteristics of connections, wood species, humidity and class of mechanical quality of the wood with the aim of designing the intervention of restoration in full compliance with the tender documents and recommendations and guide lines provided by the Consortium of experts in the year 2011–2012 [3], [8].

References

[1] Bacci, M., Bianchi, G., Campana, S. & Fichera, G. (2012) Historical and archaeological analysis of the Church of the Nativity. *Journal of Cultural Heritage*, 13, Elsevier, e5–e26.

[2] Fichera, G., Bianchi, G. & Campana, S. (2016) Archeologia dell'Architettura nella Basilica della Natività a Betlemme, pp. 1567–1589, in Olof Brandt – Vincenzo Fiocchi Nicolai (a cura di) *Acta XVI Congressus Internationalis Archaelogiae Christianae Romae* (22–28.9.2013). *Costantino e i Costantinidi. L'innovazione costantiniana, le sue radici e i suoi sviluppi.* Pontificio Istituto di Archeologia Cristiana, Città del Vaticano.

[3] Ferrara Consortium (2011) *Final Report*.

[4] Bernabei, M. & Bontadi, J. (2012) Dendrochronological analysis of timber structures of the Church of the Nativity in Bethlehem. *Journal of Cultural Heritage*, 13, Elsevier, e5–e26.

[5] Bacci, M. (2017) *The Mystic Cave: A History of the Nativity Church in Bethlehem.* Rome, Brno, Viella/Masaryk University Press.

[6] Bagatti, B. (1952) *Gli antichi edifici sacri di Betlemme in seguito agli scavi e restauri praticati dalla Custodia di Terra Santa (1948–1951).* Jerusalem, Franciscan Printing Press.

[7] Vincent, H. & Abel, F.M. (1914) Bethléem. Le sanctuaire de la Nativité. Paris, Gabalda.

[8] Ferrara Consortium (2012) *Tender Document*.

3.2 Morphologic and structural characteristics of the roof

C. Alessandri, N. Macchioni, M. Mannucci, M. Martinelli and B. Pizzo

The wooden trusses that form the main bearing structure of the roof can be divided into four main groups, corresponding to four different typologies: (a) trusses of nave, transept and apse, (b) end trusses, (c) trusses of the central area and (d) trusses of aisles and corners.

3.2.1 Trusses of nave, transept and apse

The trusses of nave, transept and apse (Figures 3.2.1, 3.2.2) are triangular trusses with two rafters, one horizontal tie beam, one vertical king post, two couples of struts connected on one side to the king post and on the other side to the rafters, but not directly, being present an element per rafter, nailed to it and called "sleeper" from here on. In particular, the trusses of nave and apse are placed in correspondence of the columns which divide the nave from the aisles and the apse from the corner areas.

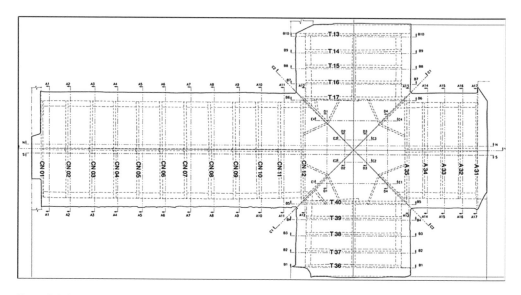

Figure 3.2.1 Trusses of nave, transept and aisles.

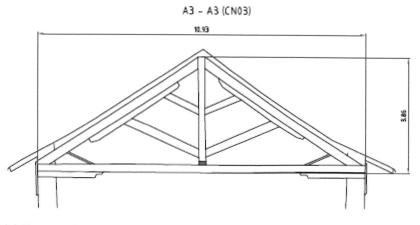

Figure 3.2.2 Truss typology in nave, transept and apse.

Figure 3.2.3 Cantilever.

The presence of double struts makes these trusses very peculiar, together with the high slope of the rafters. The span covered by the trusses varies between 9.36 m and 10.79 m (measured between the internal surfaces of the walls), the height of the trusses is contained between 3.69 m and 4.33 m, and the distance between each truss and the other is approximately 2.60 m.

Therefore, these trusses have a height/span ratio much higher than the traditional Italian trusses characterized by a low slope. The average slope is approximately equal to 80% with an angle between rafter and tie beam of about 38°.

The height of these trusses, and therefore their slope, is obviously related to the height of the trusses that cover the central part, which in turn is due to the mechanical considerations explained in detail below. Surely, if the height of the latter had been lower, the thrust transferred by their rafters to the corners of the perimeter ring (see below) would have been much higher and more difficult to balance. Hence the need for a greater slope in the trusses of nave and transept.

Two wooden cantilevers protruding from the walls support each tie beam for a variable length from 135 to 150 cm at both ends of each truss (Figure 3.2.3).

Each cantilever is connected to the wall by a horizontal iron, handmade tie rod (Figure 3.2.3) nailed to timber on one side of the cantilever and ending on the wall side with a ring containing a vertical iron wedge. In most cases these tie rods are placed on the opposite sides of each truss and are clearly identifiable from the outside through the vertical wedges (Figure 3.2.4).

They usually prevent trusses from sliding on top of the walls and reduce the horizontal relative displacements of the opposite walls in the case of seismic actions in the orthogonal direction. Rafters, tie beams, cantilevers and additional wedges between rafters and tie beams (often used to compensate for the reduced length of the rafter; Figure 3.2.5) are all joined by long nails (always ancient, forged with quadrangular section and tapered stem), often riveted at the intrados of the tie beam (Figures 3.2.6 a, b).

The ends of the cantilevers are variously worked, sometimes even with triple curved profiles (Figures 3.2.3, 3.2.5). The tie beam–rafter joint is a notched joint (rarely visible, more

Figure 3.2.4 External connection cantilever-masonry wall.

Figure 3.2.5 Nailed connection among cantilever, tie beam, rafter and rafter-tie beam wedge.

frequently hidden into the wall) and strengthened by one or two iron nails from the extrados of the rafter to the tie beam (Figure 3.2.7).

Wooden boards (Figures 3.2.8, 3.2.9) cover the tie beams of the nave from CN2 to CN11 both laterally and at the intrados, sometimes even at the extrados, in order to hide irregularities and size reduction of the beams.

a)   b)

Figures 3.2.6 a, b Riveted nails.

Figure 3.2.7 Partial notched joint between rafter and tie beam.

Figures 3.2.8 Boards covering the tie beams.

Figures 3.2.9 Boards covering the tie beams.

They are simple (sometimes double) layers nailed to the tie beams and made of low-quality wood, mostly softwood. Although the way the various truss elements are carved is generally the same, some trusses may differ from the others for some construction details. For instance, in some trusses some additional elements (as mentioned previously) are interposed between the rafter lower end and the tie beam (Figure 3.2.5).

Probably the lack of timber elements long enough to cover the whole hypothetical length of rafters forced the carpenters to insert such short horizontal elements at the extrados of the tie beam in order to complete the joint at the truss end. Moreover, in all the trusses, unlike what happens in classical Italian trusses, the king post is connected to the tie beam by means of a nailed half-lap T-joint or wedged dovetail lap joint (Figures 3.2.10 a, b and 3.2.11). The first typology is used in the nave when king post and tie beam are slightly out of the truss

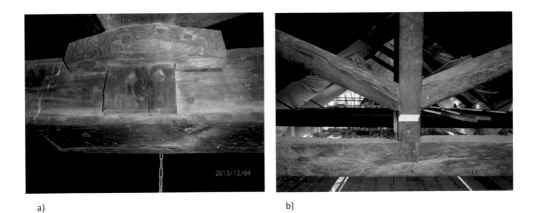

a)   b)

Figures 3.2.10 a, b Example of a half-lap T-joint on the nave and the wedged dovetail lap joint in CN12.

Figure 3.2.11 Example of a wedged dovetail lap joint on the transept.

plane (probably due to either a small deviation or a reduction in size of the tie beam). The second typology is used when both king post and tie beam remain in the truss plane (this occurs in truss CN12 of the nave and in all trusses of both transept and apse).

Such a difference might depend on the different quality and cross section dimensions of the tie beams used (both greater in the transept tie beams) as well as on the different time when such trusses were built. Because of that, the walkway in the nave, made of a single wooden board, rests directly on the extrados of the tie beam on one side and indirectly on the other side thanks to a small wooden cantilever nailed to the tie beam (Figures 3.2.12 a, b). On the contrary, the walkway in the transept is composed of two wooden beams resting on the extrados of the tie beam at both sides of the king post and covered by a decking board above and a closing board below (Figures 3.2.13 a, b).

Moreover, the beams of the nave walkway are clearly recovery elements, mostly of cedar, presumably sleepers of rafters because of the saw-worked ends, notches for strut head hosting and holes for nails on the ends (Figures 3.2.14 a, b).

a)   b)

Figures 3.2.12 a, b The walkway on the nave.

a)   b)

Figures 3.2.13 a, b The walkway on the transept trusses.

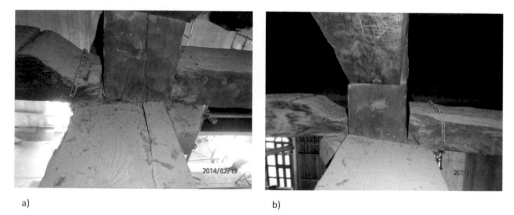

a)   b)

Figures 3.2.14 a, b Evidence of recovery elements in the nave walkway.

It is worth noting the presence of a continuous double course of wooden elements, similar to wood lacings, under the ends of the trusses of nave, transept and apse (Figure 3.2.15), whereas their presence is still uncertain in the end walls with the tympana.

These wood lacings are clearly visible on the outside (Figure 3.2.16), whereas inside they remain hidden under the plaster. They are made up of elements of various lengths, generally head-joined with half-lap wood joints (on horizontal or vertical plane) reinforced by nails or brackets nailed laterally or in the extrados (Figures 3.2.17, 3.2.18, 3.2.19, 3.2.20 a, b).

In the crossing area, at the intersection of nave and transept, the joints between the double wooden lacing elements become more complex, e.g. cross half-lap joints and corner half-lap joints with double dovetail. Moreover, some local analyses carried out at the north-west and

Figure 3.2.15 Double wooden lacings.

Figure 3.2.16 Double wooden lacings from the outside.

Figures 3.2.17, 3.2.18, 3.2.19 Double wooden lacings from the outside: different joints.

a)   b)

Figures 3.2.20 a, b Double wooden lacings from the inside: different joints (on the right (b), an iron
          bracket is also present and covered by plaster).

south-west corners of the nave (the façade of the church) highlighted the presence of at least
an inner wooden lacing element joined to the ones of the nave with a corner cross half-lap
joint. This suggests that such a double course of wooden lacings, at least originally, could
continue throughout the upper perimeter of the church (Figure 3.2.21) so as to form a sort of
reinforcement ring for the upper part of the walls.

The use of similar double wooden lacings, running along the whole perimeter of a building, belongs to a very ancient constructive tradition of the Eastern Mediterranean Countries,
such as Greece and Turkey, both frequently subject to seismic events. In the church, they
guarantee a sufficiently regular support for the trusses and a more uniform distribution on
top of the walls of the vertical loads transferred by the trusses. Consequently, even the friction forces arising between wooden lacings and masonry in the case of seismic events are
better distributed at the interface and can be, therefore, more effective in reducing a possible
horizontal sliding of the roof, as well as differential relative displacements between opposite
walls. The considerable stiffness of the whole roof, already guaranteed by the great amount
of nailed connections among its various components, is also increased by the nailed connections between wooden lacings and truss ends (Figure 3.2.22).

Near the perimeter wall there is a change in the pitch slope. Here the purlins from parallel
to the walls become orthogonal to them so that, jutting beyond the wall, they can support the
eave board (Figures 3.2.23, 3.2.24).

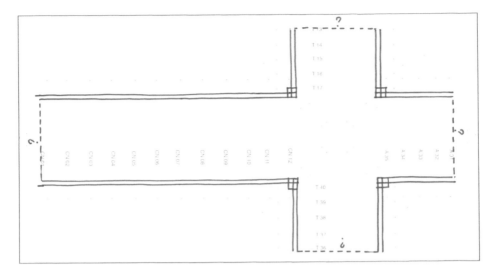

Figure 3.2.21 Double wooden lacings along the whole perimeter of the upper level.

Figure 3.2.22 Nailed connection between truss end and wooden lacings.

Figure 3.2.23 Purlins in the eave area.

In the upper part, they rest on the wooden lacings placed on top of the walls. The purlins of the eave are nailed to the nearest purlins of the nave at the upper end and to the eave board at the lower end. They are mostly made up of oak, occasionally softwoods (larch/cedar). Consequently, also the deck boards change orientation in the eave area: from orthogonal to the walls they become parallel to them (Figure 3.2.25).

Figure 3.2.24 Purlins and eave board.

Figure 3.2.25 Boards of the roof coverings.

They are simple boards, juxtaposed and nailed to the rafters, mostly cedar, probably obtained by sawing the elements recovered from the demolition of the previous roofing structure.

Rafters and king posts are connected through notched joints on the king post, reinforced by 1–2 iron nails from the intrados of the rafters to the king posts (Figure 3.2.26).

Mostly everywhere the struts are joined to rafters and king posts through notched joints (Figures 3.2.27, 3.2.28).

Figure 3.2.26 Notched joint on the king post.

Figure 3.2.27 Notched joint struts – king post.

Figure 3.2.28 Notched joint struts – sleeper.

Figure 3.2.29 CN 01 – Truss on the west tympanum.

3.2.2   End trusses

Only the trusses in correspondence of the tympana of nave and transept are simpler because they are without king post, struts, sleepers and iron brackets (present in all the other trusses). Unlike the others, they have a collar beam and bear no other load except their own self-weight because the roof purlins above them lean directly upon the top of the tympanum. They have only a decorative function that consists in hiding the insertion of the purlins directly into the tympanum. Nevertheless, like the others, they guarantee an effective connection between the supporting walls (Figure 3.2.29).

3.2.3   Trusses of the central area

Very complex is the timber structure of the central area (Figures 3.2.30, 3.2.31) at the crossing of nave and transept. Two big diagonal trusses (in light blue in Figure 3.2.32), from the S-W corner to the N-E corner (n° 1 in Figure 3.2.33, or CA22 – CA24 in Figure 3.2.34) and from the N-W corner to the S-E corner (n° 2 in Figure 3.2.33, or CA23 – CA25 in Figure 3.2.34), cross that area, having, at the right centre, a common big king post (C in Figure 3.2.33), connected to each rafter by a couple of struts.

Figure 3.2.30 Trusses in the crossing area.

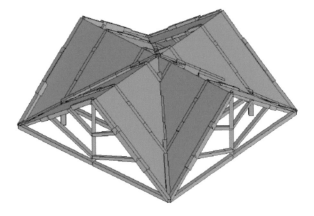

Figure 3.2.31 3D view of the crossing area.

Figure 3.2.32 Crossing area.

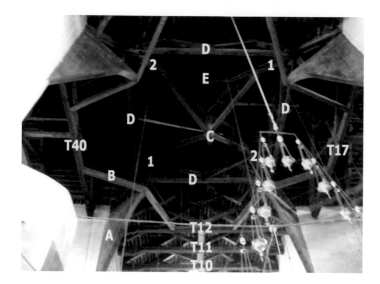

Figure 3.2.33 Crossing area.

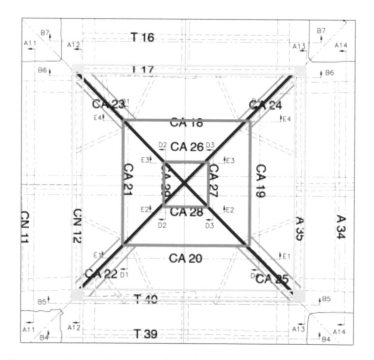

Figure 3.2.34 Plan view of the main structural components of the roof covering the central area: in yellow the four perimeter trusses belonging to nave, transept and apse; in red the two diagonal trusses; in green the smaller trusses and in blue the corner cantilevers with the "incomplete tie beams".

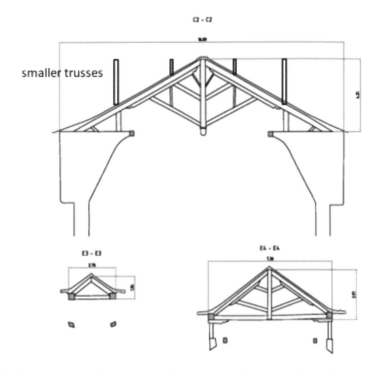

Figure 3.2.35 Diagonal truss with corner cantilevers and smaller trusses (in red).

Further details of the whole structure are shown in Figure 3.2.35.

The diagonal trusses do not have tie beams, or at least ordinary tie beams connecting the lower ends of the rafters. Therefore, it is natural to ask how the thrust exerted by the rafters of these two large trusses can be absorbed. As can be seen in Figure 3.2.34, several structural elements meet at the corners of this area: the diagonal trusses CA23 – CA25 and CA22 – CA24, the T17 and T40 trusses of the transept, the CN12 truss of the nave and the A35 truss of the apse. Moreover, at each corner there is a horizontal element in the diagonal direction, from now on called "incomplete tie beam" (Figure 3.2.34), supported by a large triangular-shaped cantilever structure, anchored to the wall and connected to the rafter of the diagonal truss by a couple of iron brackets (Figures 3.2.36, 3.2.37). The inner end of the "incomplete tie beam" is connected to the tie beams of the four perimeter trusses by two other horizontal inclined elements, nailed to the adjacent tie beams, symmetrical to each other and probably aimed at stabilizing the triangular cantilever.

The cantilevers, connected to the lower ends of the diagonal trusses, also offer an indirect support to four smaller trusses, denoted by D in Figure 3.2.33 and as CA18, CA19, CA20, CA21 in Figure 3.2.34 and connected to the cantilever by a short vertical wooden pillar (Figure 3.2.36). The main structure of these cantilevers is formed by the upper horizontal element mentioned above, one vertical element inserted in the wall and two inclined struts, all nailed together. In particular, the horizontal element, covered on both sides by boards and connected to the truss rafter by two iron brackets, is generally composed of some overlapping and nailed pieces, some of them half-lap jointed (Figure 3.2.38).

Figure 3.2.36 View of the "incomplete tie beam" system.

Figure 3.2.37 Corner triangular cantilever.

Figure 3.2.38 Upper horizontal element.

Some horizontal boards, nailed to the struts, support the external vertical boards that cover the whole cantilever on both sides (Figures 3.2.39, 3.2.40).

In each of the four corners in which the longitudinal body of the church intersects the transept, double wooden lacings, cantilevers and tie beams of the perimeter trusses of the central area and the "incomplete tie beams" of each diagonal truss extend out of the masonry (Figures 3.2.41 a, b) and join to each other in various ways, through corner half-lap joints, mitred half-lap joint or mixed joints, solidarized by nails and nailed brackets (Figures 3.2.42, 3.2.43).

Each protruding part is supported by a vertical reinforcement made of the same stone of the church, clearly visible on the outside and as tall as the parts of wall that rise above the aisles (Figures 3.2.44 a, b).

Figure 3.2.39 Struts, horizontal and vertical boards.

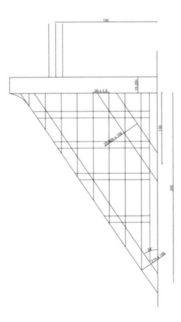

Figure 3.2.40 Struts, horizontal and vertical boards.

Figures 3.2.41 a, b Corner elements protruding outside: (a) view from the inside and (b) from the outside.

Figure 3.2.42 Intersection of T17, CA23, double wooden lacings and triangular cantilever.

This reinforcement is made by extending outwards the corner stones of the walls, some of which have the L shape already encountered in other parts of the church and useful to ensure a better anchorage to the rest of the building. As this reinforcement is part of the original structure of the church, it is reasonable to assume that even the original roof used this reinforcement with the same purpose. Therefore, it is possible to think that the original roof had

Figure 3.2.43 Connection of the cantilever vertical element with masonry and double wooden lacings.

a)   b)

Figures 3.2.44 a, b External corner reinforcement.

the same structure as the current one, or at least a very similar structure, which was preserved in subsequent restoration interventions.

The good connection, in correspondence of the corners, among the perimeter trusses (CN12, T17, T40, A35) and the diagonal trusses (CA22 – CA24, CA23 – CA25) allows us to suppose that the horizontal component of the thrust exerted by each diagonal truss is absorbed by the tie beams of the perimeter trusses. In this way, the four perimeter tie beams would behave like a closed ring subject to tensions also coming from the diagonal trusses (Figure 3.2.34).

Because of the high concentration of loads transferred to the corners by all the structural elements of the central area, the corner wooden joints are tightened by a sort of ring formed by iron brackets nailed to the protruding wooden parts (Figure 3.2.45).

Four smaller trusses parallel to the trusses of nave, transept and apse (D in Figure 3.2.33, in red in Figure 3.2.32, CA18, CA19, CA20 and CA21 in Figure 3.2.34) rest on the small pillars above the big cantilevers. As the ordinary trusses, they have one king post connected to the tie beam by means of a nailed half-lap T-joint. Their tie beams are joined to the rafters of the diagonal trusses with a different variant of notched joints (Figures 3.2.46 a, b), whereas their rafters are simply nailed on them.

Figure 3.2.45 Iron brackets as reinforcement of the corner wooden joints.

a)   b)

Figures 3.2.46 a, b Joint between the diagonal truss rafter and the small trusses (front (a) and rear (b) views).

Figure 3.2.47 Joint between orthogonal cantilevers above each small pillar.

Figure 3.2.48 Pillar–triangular cantilever connection.

Moreover, each couple of cantilevers are joined on top of each small pillar by a corner half-lap joint with double dovetail (Figure 3.2.47) and by a round tenon-and-mortise joint (it was seen through a small disconnection). Each pillar is end-joined to the big triangular cantilever by using inclined nails (Figure 3.2.48).

It seems that there are no direct connections between tie beams and rafters of the small trusses, as they both intersect the rafters of the diagonal trusses through carved and nailed joints (Figure 3.2.49).

A third level of very small trusses (E in Figure 3.2.33, in red in Figure 3.2.32, CA26, CA27, CA28, CA29 in Figures 3.2.34 and 3.2.35) is placed close to the head of the big central king post. Their rafters and tie beams, the only components of such trusses, join alongside the rafters of the diagonal trusses, at the same height as the struts, by means of nailed joints, whose structure was not investigated (Figure 3.2.50).

The wood species identified in the wooden structural elements of the whole upper level (mainly deciduous oak, cedar, larch) are mostly associated with the types of element. For

Figure 3.2.49 Indirect connection between tie beams and rafters of small trusses.

Figure 3.2.50 Third-level small trusses: connection between tie beam, rafter, the related rafter of the diagonal truss and the strut coming from the central king post.

example: various species in cantilevers; more often cedar in sleepers; oak always in struts, tie beams, king posts and external wood lacings; oak, larch (cedar) and occasionally beech in inner wood lacings.

Oak and cedar are generally of medium-low quality, sometimes strongly defective for grain deviations, presence of knots and ring-shakes.

3.2.4   Trusses of aisles and corners

The trusses of aisles and corners (areas placed at the intersection between apse and transept) (Figure 3.2.51) are half trusses, or "false" trusses, with identical geometrical features (Figure 3.2.52), apart from some small dimensional changes in the corners.

They are composed of three struts, rafter and tie beam generally interrupted at half span in correspondence of a short masonry pillar protruding from the wall above the colonnades of the aisles (Figure 3.2.52).

The rafters are often made of two elements joined in various ways (scarf joint, half-lap joints, with oblique or orthogonal cut, with nails and/or nailed brackets (Figures 3.2.53 a, b).

On the contrary, it is not possible to say anything about the continuity of the tie beam inside the short masonry pillar at half span. Consequently, it is not possible to say if the tie beam is able to absorb the thrust coming from the rafter, as it should. The struts are not directly connected to the rafters, but through a sleeper, as in other trusses (Figure 3.2.54).

In the aisles, only a single course of wooden lacings was found, along the external walls, generally hidden by the plaster (Figure 3.2.55). Thus, the wooden lacings were surveyed only on their surface towards the interior of the church.

The end trusses (LNN1 and LNS1, LNN12 and LNS12) have a vertical wooden support ("false king post") instead of the small masonry pillar and tie beam without the innermost element (Figure 3.2.56).

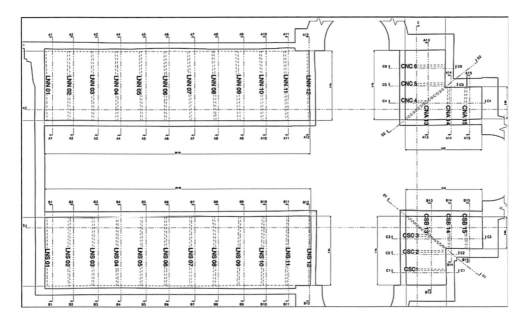

Figure 3.2.51 Plan of trusses in aisles and corners.

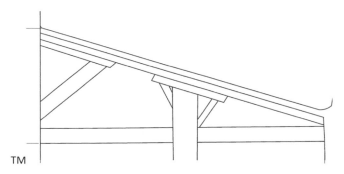

TM

Figure 3.2.52 Half truss in aisles and corners.

a)   b)

Figures 3.2.53 a, b Head joints in rafters.

Figure 3.2.54 Sleeper between rafter and struts.

Figure 3.2.55 Wooden lacing of the aisle (present on the inner surface of external walls).

Figure 3.2.56 End truss LNN1.

The purlins have irregular shape and are usually made of softwood (cedar recovered from the previous structure) and hardwood (oak). Those on the end trusses do not enter the wall, whereas the holes of the purlins of the original structure became visible during the restoration works (Figure 3.2.57).

From LNN6 to LNN12 (north wall) there are small stone cantilevers under the truss ends, added in the past in replacement of severely decayed parts of the wooden lacing elements. In some cases, a triangular wedge was also inserted between tie beam and rafter (Figure 3.2.58).

Figure 3.2.57 Holes for the purlin ends of the original structure.

Figure 3.2.58 Stone cantilever and triangular wedge between tie beam and rafter in one representative aisle truss.

As for the wood species, most of the elements are made of deciduous oak, cedar, larch. In particular, tie beams and struts are predominantly made of oak (occasionally cedar, and in this case it is recovery material), struts and sleepers often of recovery cedar (easily identifiable for the presence of special workings and holes left by nails related to previous functions/locations), wooden lacings on the south side mainly of oak and mainly of larch/cedar on the north side.

Appendix

Wood joints belonging to cultural heritage may be very complex and with variable name attribution (which often depends on local traditions and may vary accordingly, although limitedly). Thus, the present Appendix shows some schematic drawings of the typologies representing the main carpentry joints found in the church and described in the text.

For the sake of clarity, they have been grouped into (a) angle joints, (b) transversal joints, (c) corner joints and (d) longitudinal joints.

Angle joints

Joints belonging to this typology are those connecting the various elements of trusses: tie beam/rafter; rafter/king post; king post/struts; struts/sleeper (the tie beam/king post joint is described in the "transversal joints" section). Although angle joints are the most abundant ones, the basic typology is always the same, that is, the notched joint schematically represented in Figure 3.2.A.1.

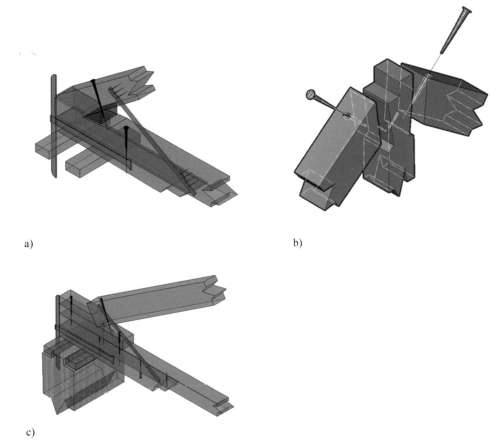

a)

b)

c)

Figure 3.2.A.1 Schematic drawings of notched joints in: (a) tie beam–rafter connections (the triangular wedge is only present in aisles); (b) rafter–king post connections; (c) tie beam–rafter connections with wedge (only present in nave and transepts).

Transversal joints

This typology is the most diversified one. It includes the joints between king post and tie beam, which are of two types: the simplest one is the half-lap T-joint (Figure 3.2.A.2), with the tie beam not notched to compensate for small irregularities, while for in-plane connections standard wedged dovetail lap joints (Figure 3.2.A.3) are used.

Transversal joints are largely adopted in the crossing area at the intersection of nave and transept. Some of the elements constituting the wood lacings on the upper walls of the roof (Figure 3.2.21) are connected by means of cross half-lap joints (Figure 3.2.A.4), in addition to the corner half-lap joints with double dovetail shown in subsection "Corner joints"

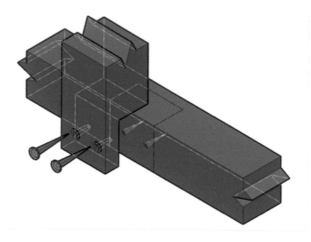

Figure 3.2.A.2 Schematic drawing of half-lap T-joints.

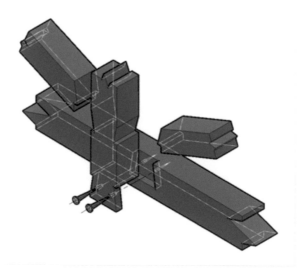

Figure 3.2.A.3 Schematic drawing of wedged dovetail lap joints.

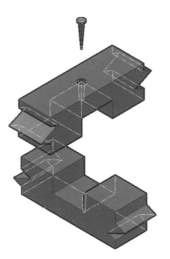

Figure 3.2.A.4 Schematic drawing of cross half-lap joints.

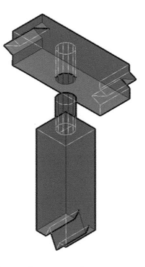

Figure 3.2.A.5 Schematic drawing of round tenon-and-mortise joints.

(Figure 3.2.A.6). Always in the crossing area, the pillars supporting the smaller trusses are connected to them by means of round tenon-and-mortise joints (Figure 3.2.A.5).

Corner joints

Joints of this typology include the corner half-lap joints with double dovetail (Figure 3.2.A.6) connecting each couple of orthogonal cantilevers above the small pillars of the crossing area, and the corner elements of the wood lacings of the roof, inside the walls (Figure 3.2.21). The elements present in the corners outside the walls are connected by means of corner half-lap joints (Figure 3.2.A.7), mitred half-lap joints (Figure 3.2.A.8) or mixed joints.

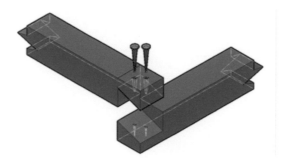

Figure 3.2.A.6 Schematic drawing of corner half-lap joints with double dovetail.

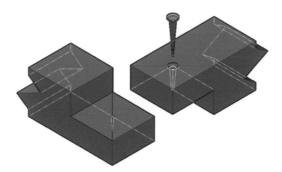

Figure 3.2.A.7 Schematic drawing of corner half-lap joints.

Figure 3.2.A.8 Schematic drawing of mitred half-lap joints.

Longitudinal joints

These joints are extensively used to connect the elements constituting the wood lacings on the upper walls of the roof. Longitudinal connections were executed by means of half-lap joints, on either the horizontal or the vertical planes, and are schematically represented in Figure 3.2.A.9.

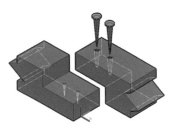

Figure 3.2.A.9 Schematic drawing of longitudinal half-lap joints.

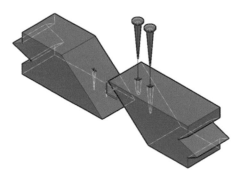

Figure 3.2.A.10 Schematic drawing of scarf joint.

Moreover, longitudinal connections are also present in the rafters of the aisle trusses. In this case, different typologies are observed such as scarf joint (Figure 3.2.A.10) and half-lap joints.

3.3   The scaffolding

C. Alessandri and M. Martinelli

The most urgent works on the roof had to start at the beginning of winter 2013. It was therefore necessary to design a scaffolding that, in addition to providing the necessary platforms for the work to be carried out from the inside, could also allow a coverage for all works that should have been done from the outside, at least during the wintertime. The scaffolding design had to take into account some conditions, already manifested in the design of the previous scaffolding built in the first 2010 diagnostic campaign [1]: all the activities usually carried out inside the church should not have undergone any variation, the floor of the church should not have been subjected to strong compressions, the columns of the scaffolding should have been positioned outside the areas comprising the underground cavities in order to avoid strong stresses on the vaults and consequent damages, the scaffolding should not have had supports in correspondence of the walls of the nave so as not to exert pressure such as to cause detachments of the wall mosaics from their support, and finally the scaffolding modules and the assembly system should have guaranteed maximum flexibility and therefore the possibility of making changes in the scaffolding according to the changing needs of work.

The well-known MP Pilosio multi-directional scaffolding system was chosen with tubular elements and steel profiles S235J0 (f_{jk} = 235 N/mm^2 – yield stress; f_{tk} = 360 N/mm^2 – failure stress) protected by hot galvanising (average thickness 55 μ). Other mechanical properties of the materials used can be found in [2]. The design of the scaffolding for the church and the corresponding structural calculations were entrusted to CMW Engineering srl, Florence, Italy (responsible for structural analyses: Eng. Alessandro Incerpi). Since the first diagnostic analyses and the first restoration interventions concerned the roof of the upper level, the first scaffolding to be designed and built was that related to the nave, also including the presbytery and the transept. Instead, the scaffolding for analyses and the interventions in the wooden structures of the aisles was built later with variable extensions according to the needs. The structure of the scaffolding was designed and calculated according to European and Italian Standards [3–6].

Work began at the beginning of winter 2013. It was therefore necessary to create an external temporary cover that would allow workers to work from outside, at least for all the wintertime. This external metal structure covered the roof in correspondence of the first four spans, starting from the façade of the church, and was supported only by the columns of the internal scaffolding, to which it was connected by horizontal metal elements passing through the window openings. Externally, the structure was connected to the north and south walls (more precisely to the additional walls built in the Crusader period above the perimeter walls) by means of inclined truss beams following to the slope of the pitches, anchored on one side to the masonry and on the other to the tubular elements of the main structure, as clearly shown in sketches 3.3.1 and 3.3.2 and in Figures 3.3.3, and 3.3.4. It should be noted that internally the structure was not clamped to the ground, for obvious reasons of respect for pre-existence. Therefore, those external connections had only to prevent or limit possible horizontal oscillations of the entire scaffolding due to the wind acting on the external surfaces, without transferring vertical loads to the walls. Other constraints to prevent horizontal movements of the structure were in correspondence of the façade of the church (Figure 3.3.5) and at both sides of the window openings (Figure 3.3.6).

The tarpaulin used to cover the external structure, provided by Ortona srl, Livorno, Italy, was a polyester PVC membrane in modular sheets 12.50 m long and 3 to 5 m wide, fastened to the tubular elements by means of elastic belts passing across eyelets (Figures 3.3.7, 3.3.8).

Inside the nave, the scaffolding, represented schematically by the longitudinal and vertical sections in Figure 3.3.11, left the whole central part of the nave completely free (Figures 3.3.12, 3.3.13) and had two floors, as shown more clearly in Figure 3.3.2: the first as an access area from the windows, deposit of work tools, samples, meeting area, restoration of the wall mosaics at the windows level, etc. (Figure 3.3.14); the second was reserved for interventions on the trusses of the upper level (Figure 3.3.15).

The external cover was limited only to this first part of the roof and removed, as soon as the works in this part were completed; all the rest of the roof did not need any external cover

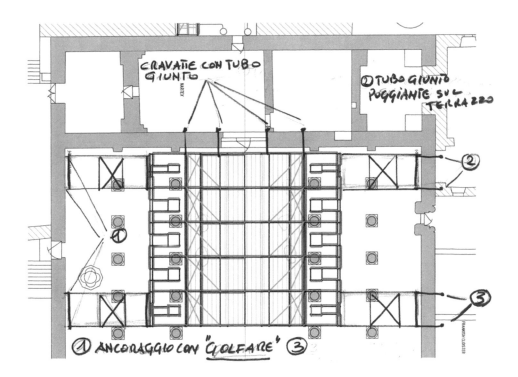

Figure 3.3.1 Plan view of the external scaffolding.

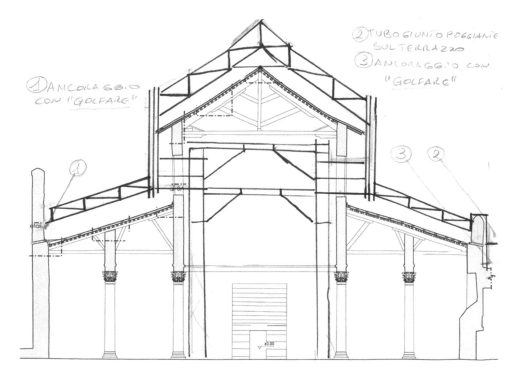

Figure 3.3.2 Vertical section of the first part of the scaffolding.

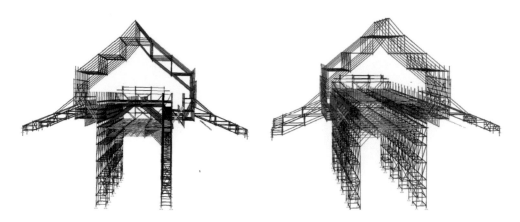

Figure 3.3.3 Perspective views of the whole scaffolding.

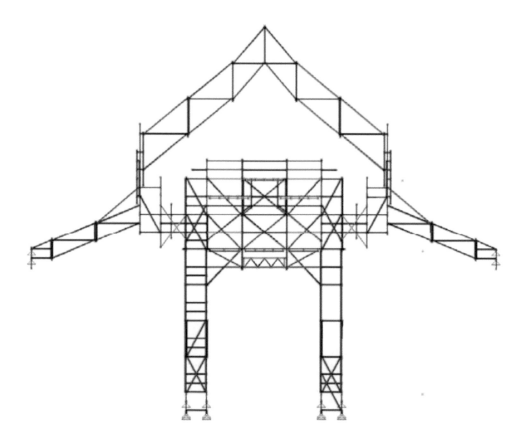

Figure 3.3.4 Vertical section of the first part of the scaffolding.

Figure 3.3.5 Connection with the façade of the church.

Figure 3.3.6 Connection with the window openings.

Figure 3.3.7 Tarpaulin and external connections.

Figure 3.3.8 Covered scaffolding.

Figure 3.3.9 Scaffolding above the roof.

Figure 3.3.10 Scaffolding coming out of the windows.

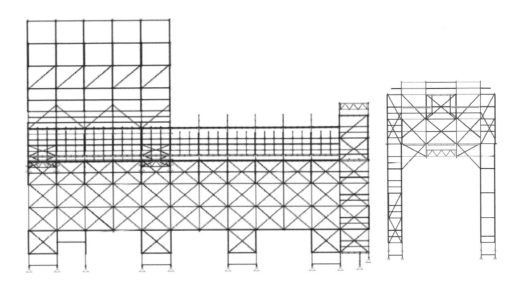

Figure 3.3.11 Longitudinal and transverse section of the scaffolding in the nave.

Figure 3.3.12 Free area at ground level.

Figure 3.3.13 Free area at ground level.

Figure 3.3.14 First floor.

Figure 3.3.15 Second floor.

Figure 3.3.16 External scaffolding stopped at the eave level.

as the interventions could be carried out during the good season. Therefore, externally, the scaffolding coming out of the windows was stopped at the level of the eaves (Figure 3.3.16) to allow specialized personnel to remove the existing lead sheets, restore the old wooden plank and proceed with the laying of the new roofing.

Internally, between the base of the columns of the scaffolding and the floor, a wooden nut was inserted, sufficiently high and wide to diffuse on wider surfaces the vertical load coming from the scaffolding and to subject the stone slabs of the flooring to acceptable compressive stresses (Figures 3.3.17, 3.3.18).

It is worth remembering that in the nave the columns of the scaffolding are all arranged outside the area corresponding to the floor mosaics of the first church of Constantine, at the time of the works still partially hidden under the floor.

Analogous criteria were followed for the design and construction of the scaffolding in the transept and central area (Figures 3.3.19, 3.3.20), which both had to remain usable at ground level for daily liturgies and access to the grotto.

Also in this case the control of the pressure exerted on the base had to be very accurate given the presence of the grotto of the Nativity in the central area and the extremely complex and articulated grottoes of Saint Jerome on the north side (Figure 3.3.21).

As soon as the restoration of the roof was completed, the upper level of the scaffolding was dismantled. Only the first level remained until the restorations of the angels was completed. Then even this level was dismantled and only a double walkway on both sides

Figure 3.3.17 Wooden nut.

Figure 3.3.18 Wooden Nut.

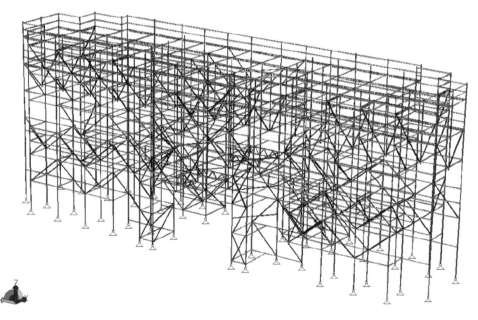

Figure 3.3.19 3D view of the scaffolding in central area and transept.

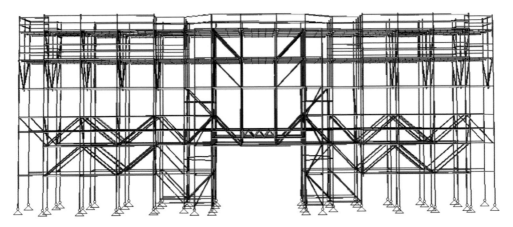

Figure 3.3.20 Perspective view of the scaffolding in central area and transept.

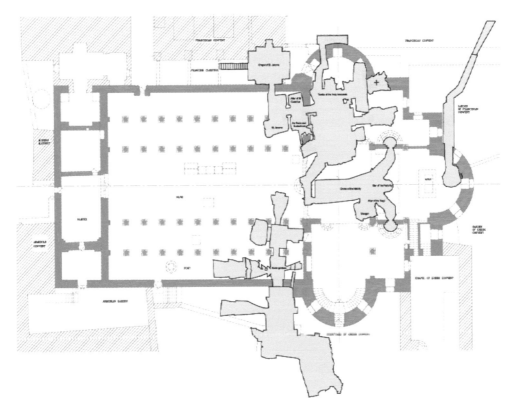

Figure 3.3.21 Grottoes in central area and transept.

Figure 3.3.22 Side walkways in the nave.

Figure 3.3.23 Side walkways in the nave.

of nave and transept was left in order to complete the restoration of the lower wall mosaics (Figures 3.3.22, 3.3.23).

References

[1]   Ferrara Consortium (2011) *Final Report*.
[2]   Autorizzazione Ministeriale n. 20034/OM.4-25/01/2001.
[3]   EN 1991 – Eurocode 1: *Actions on structures*.
[4]   EN 1993 – Eurocode 3: *Design of steel structures*.
[5]   D.M. LL.PP. 18/01/2008: *Norme tecniche sulle costruzioni*.
[6]   Circolare n. 617, 02/02/2009: *Istruzioni per l'applicazione delle Nuove Norme Tecniche per le Costruzioni di cui al D.M. 14/01/2008*.

3.4 State of conservation of the wooden structures

N. Macchioni, M. Mannucci, M. Martinelli and B. Pizzo

In order to allow a cataloguing and an easy recognition of the trusses, numbering was carried out according to their belonging to the different parts of the church as follows (Figure 3.4.1):

Nave:	CN 01 to CN 12 from the façade to the crossing area
North transept:	T 13 to T 17 from the north wall to the crossing area
South transept:	T 36 to T 40 from the south wall to the crossing area
Apse:	A 31 to A 35 from the east wall to the crossing area

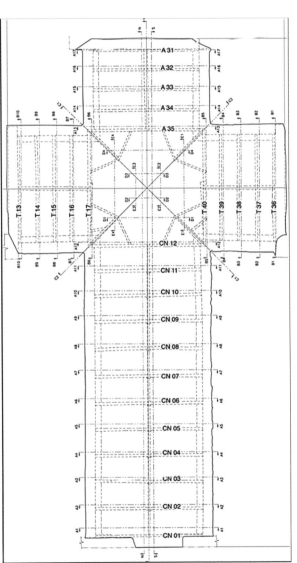

Figure 3.4.1 Trusses of nave, transept and apse.

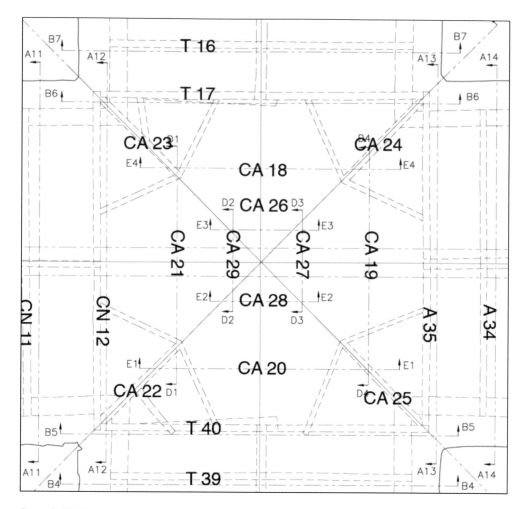

Figure 3.4.2 Trusses in the crossing area.

In the crossing area (Figure 3.4.2) the four small trusses are from CA 18 to CA 21 going clockwise from the north side (Figure 3.4.2) and denoted by D in Figure 3.4.3. Instead of numbering the two diagonal trusses (1 and 2 in Figure 3.4.3), it was decided to number the corresponding four rafters: from CA 22 to CA 25 going clockwise from the SW corner (Figure 3.4.2); the same number was also given to the corresponding upper and lower struts. The smallest four trusses (E in Figure 3.4.3) are numbered from CA 26 to CA 29 going clockwise from the north side (Figure 3.4.2). The element 30 is the big central king post (C in Figure 3.4.3).

In the aisles the truss numbering follows the same scheme as for the nave. So they are divided into LNN (lateral nave north or north aisle) and LNS (lateral nave south or south aisle) and numbered from 1 to 12 starting from the façade (Figure 3.4.4). Existing numbering was used for the trusses of the corners between transept and nave. The letters CN (corner

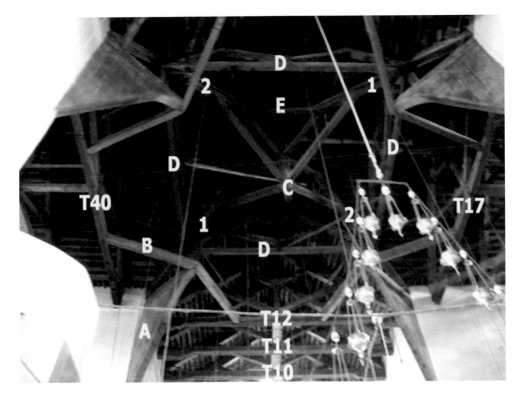

Figure 3.4.3 Trusses in the crossing area.

north) and CS (corner south) distinguish the trusses of the north corner from those of the south corner (Figure 3.4.4).

The timber grading was performed according to the Italian Standard UNI 11119:2004 (grading table at page 4) [1] that identifies three classes (here expressed as grades I, II, III) corresponding respectively to the following quality levels: better quality, average quality, lower quality. The timber elements having a very low structural quality are marked as N.S. (not suitable for mechanical purposes); The table at the end of the Standard suggests the maximum stresses allowed, within the admissible stress method, and the average bending MOE (modulus of elasticity) of each on-site grade of main timber species, applicable for wood moisture content = 12%. For N.S. elements there are no mechanical profiles suggested. Elements marked as N.G. are not gradable due to contingent situations, such as covered elements or composed elements. As an example, Table 3.4.1 shows the timber grading referred to truss CN 02.

The extension of the decay was measured by means of resistographic drillings carried out with an IML Resi B400 drilling device (see www.iml.de), electronically controlled and able to reach a depth of 40 cm with a resolution of 0.02 mm. The decay was quantified and represented by using four percentage ranges, i.e. 0–25%, 25–50%, 50–75% and 75–100%, corresponding to states of increasing decay and denoted by four different colors in the shop drawings.

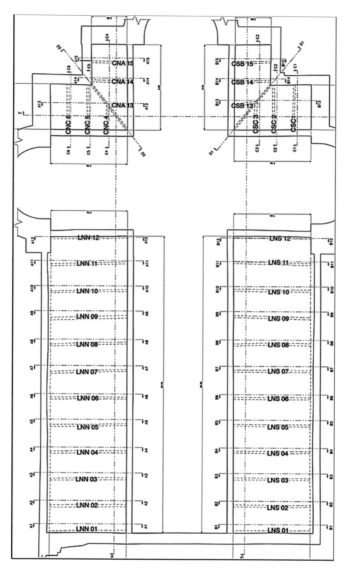

Figure 3.4.4 Trusses of aisles and corners.

Table 3.4.1 Timber grading – truss CN 02

Truss	Tie beam		Rafter N		Rafter S		Strut Upper		Strut Lower		Sleeper		King post
	N	S	Beam	Apex	Beam	Apex	N	S	N	S	N	S	
02 CN	20	18×18	23×23	18×22	24×29	24×24	22×27	21×21	22×27	21×28	16		19×20

On-site structural grading:

Tie Beam	North Rafter	South Rafter	King Post	Strut Upper		Strut Lower	
N.G.	III	N.S.	I	N	S	N	S
				N.S.	II	I	I

In the case of decay, the previously described structural grading was applied with reference to the undamaged part of the beam. Figure 3.4.5 shows, for instance, the state of decay for truss CN 04 in the nave according to the criterion mentioned above.

In the structural elements of the upper level (trusses of nave, transept, apse and central area), the decay was mostly due to rot, from moderate to very severe (Figures 3.4.6, 3.4.7). A moderate decay caused by anobids/cerambycidae, mostly limited to sapwood, was surveyed occasionally and in negligible percentages, whereas decay due to termites was absent.

In aisles and corners the decay was mainly concentrated in the truss ends of the external walls, and it was mainly due to fungi and underground termites, only occasionally and in negligible percentages due to anobids (Figure 3.4.8). A high percentage of decay was found even in the inner wooden lacing elements (Figure 3.4.9).

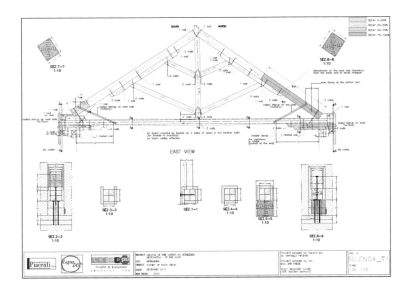

Figure 3.4.5 State of decay in truss CN 04.

Figure 3.4.6 Decay in the south rafter of CN 09.

Figure 3.4.7 Decay in the south rafter of CN 11.

Figure 3.4.8 Decay in the truss end.

Figure 3.4.9 Decay in the inner lacing elements.

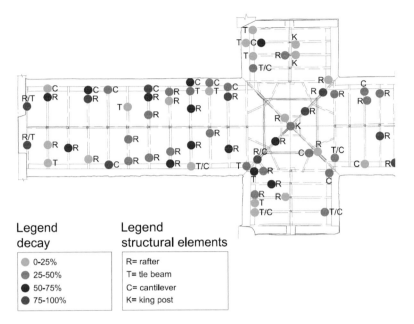

Figure 3.4.10 Decayed trusses of the upper level.

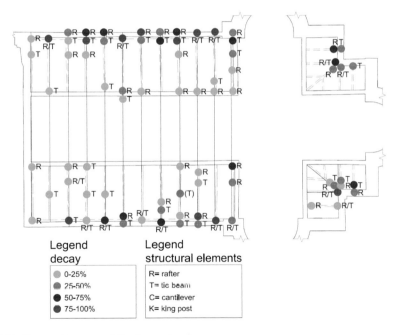

Figure 3.4.11 Decayed trusses of the lower level.

Figures 3.4.10 and 3.4.11 show the localization of the decay in the trusses of the upper and lower level respectively. In each decayed truss the percentage of decay and the damaged truss elements are shown.

Reference

[1]   UNI 11119:2004 – *Beni culturali – Manufatti lignei – Strutture portanti degli edifici – Ispezione in situ per la diagnosi degli elementi in opera (Wooden artefacts – Load-bearing structures – on-site inspections for the diagnosis of timber members)*.

3.5   Wooden species and dating of the wooden structural components

M. Bernabei, J. Bontadi, N. Macchioni and M. Mannucci

The species, determined per each element, were identified through both macroscopic evaluation and microscopic analysis. The microscopic identification was carried out by analyzing in the Laboratory some samples taken on site. The most important wood species identified in the trusses are oak (*Quercus* sp.), Lebanon cedar (*Cedrus libani* L.) and European larch (*Larix decidua* Mill.). Some sparse elements of cypress (*Cupressus sempervirens* L.) and pine (*Pinus sylvestris* L. or *Pinus nigra* Arn.) are also present. Among the listed species only cedar and cypress can be considered native, whereas more probably larch has an alpine origin and oak is probably European or Turkish. Pine could come from both Europe and Turkey. Their distribution within the whole timber structure of the roof is the final output of the different phases of construction, dismantling, reconstruction and maintenance of the roof over the centuries. It was extremely difficult to associate each species with each phase of the roof life because the historical evidence available, including the so called "firmans", or *ferman* in Turkish, are very vague and talk about reconstruction and maintenance interventions without specifying in detail the types of wood used and the localization of the intervention. However, the identification of the different wood species used allows saying that the principal elements of the trusses are made of oak and larch, and only smaller and secondary elements are made of cedar. It also allows making some hypotheses about their provenance and, in some cases, possible temporal connections with the historical records. An exact dating of the roof timber elements could only be accomplished by means of the dendrochronological analysis, which, however, would have required times and costs well above those afforded. Anyway, the dendrochronological and radiocarbon analysis (C14) carried out on some samples in the 2010 campaign, and more recently in 2013–2014, provided some interesting results. It is worth noting that, whereas the dendrochronological analysis allows, in the most favourable cases, the year when the tree was cut down, the C14 analysis provides a time range, depending on the features of the samples, within which the tree was cut down. According to such analyses most of the secondary components of the trusses in the aisles (cantilevers, sleepers, struts) and a few components of the trusses of the crossing area (tie beams) are made of larch cut between 1440 and 1460 A.D. This proves that these components were brought to the site, during the restoration works started in 1479, from the Eastern Alps belonging, at that time, to the Republic of Venice that had close trade relations with the Middle East. We should not be surprised by the long period elapsed between the date when the trees were cut down and the date of the actual interventions. Such a delay was probably due to the diplomatic obstacles among the subjects involved and to the logistic problems of transportation and processing. Most of the remaining structure of the church is made of oak (tie beams, rafters, struts, some cantilevers, curbs along the walls, purlins) dating back to 1848, the time when some reconstruction works were carried out after the 1834 earthquake. The dendro-chronological analysis proved the Turkish provenance of the trees used, so underlining the analogy with some interventions made in the same period and with the same wooden species in Hagia Sophia in Constantinople.

Some components of the secondary roof structures (sleepers, struts, curbs, purlins) and all wooden architraves are made of cedar. In particular, it is possible to identify three groups of cedar structural elements:

1 Architraves above the colonnades (650 A.D. ± 60 years), dating back to the time of Justinian's reconstruction. They lead us to think that the whole timber structure was originally made of cedar.
2 Some structural components of the roof (1164 ± 76 years), dating back to the time of the Crusades. This period is characterized by important works carried out inside the church (paintings on columns, main altar in the grotto, wall mosaics) and minor restoration works on the roof, like localized repairs made by using wooden elements already present in the roof at that time.
3 Some elements of the roof (1591 ± 75 years), the presence of which is justified by minor restoration works carried out by Greek carpenters in 1670–71, as reported in some historical documents. Because they used wood brought from Constantinople, it is reasonable to assume that these cedar elements were part of the same timber delivery.

In the end, it is worth noting that the roof structure at the beginning of the restoration works was undoubtedly the result of several interventions over time, consisting of repairs/replacements/reconstructions with a frequent re-use of parts of the previous structure. However, despite the information gathered and for the reasons mentioned above, it was impossible to localize the various interventions made over the centuries. For this reason, it was decided to preserve as much as possible the existing wooden components of the roof.

There are two *Cedrus* species in the Mediterranean area: *Cedrus atlantica* (Atlas cedar), which takes its name from the Atlas mountain in Morocco and Algeria, and *Cedrus libani* (Lebanon cedar). From the anatomical point of view the wood produced by the species cannot be differentiated. In this case the species was identified by the proximity of the Lebanon mountains to the building yard.

More difficult was to identify the date of previous interventions on the lead sheets of the covering structure. In this case the interventions of replacement were carried out over time with even greater frequency and, most of the time, without leaving any historic record. But unlike the wooden structures, most of the lead sheets were seriously damaged and far below any acceptable standard of reliability, so as to make advisable their total replacement. Before their removal, however, all sheets were marked, and a detailed plan of their location within the entire covering was drawn. Moreover, after the removal, the sheets and most of the nails used for connections were stored in appropriate spaces.

The average thickness of the lead sheets was 3.5 mm, but in some parts it was nearly 2 cm because of deformations produced by the high temperatures that can be registered in summer months. The existing roof covering was made of an insulating layer of earth and straw, lead sheets and layers of bitumen and canvas impregnated of bitumen, which were not suitable any more to guarantee the waterproofing of the roof.

3.6   Interventions on trusses, purlins and boards

C. Alessandri, V. Mallardo, M. Martinelli and B. Pizzo

The safeguard of historical buildings is always a difficult task, because following the same models and rules usually adopted for new constructions can lead to the choice of useless or even harmful structural interventions and, in certain cases, it can even mislead about the real safety level of the structures. It is nowadays largely accepted that the interventions on historical buildings cannot be so invasive as to modify permanently their cultural value and, generally speaking, their structural behavior. In the case of the Nativity church of Bethlehem, this approach was also adopted for the structural elements of the whole roof. This measure allowed preserving the technologies and the construction rules adopted in the past as well as the majority of the original materials the trusses, the purlins and the boards are made of. It also allowed improving the structural safety levels while respecting the architectural, historical and artistic peculiarities of the church. Therefore, the interventions carried out were at the same time consistent with the general Principles of Conservation and Restoration and with the structural safety requirements of the National and International Codes (Italian Technical Standards and European Codes for Buildings, compatible with the main Charts of Conservation and Restoration and the UNESCO principles, were mostly followed).

More in detail, the critical zones where a structural intervention was actually needed were identified based on the diagnostic survey carried out as described in Section 3.4. The choice of the specific intervention was, of course, influenced by the safety level of the damaged structural element. The structural checks were carried out with reference to procedures respectful of the European [1][2] and Italian Technical Standards [3], and they were implemented in the framework of both the "Admissible Stresses" and of the "Ultimate Limit State". Such checks made it possible to define, with reasonable approximation, the zones/parts of the roof structure characterized by significant damage and for which structural interventions were actually needed. Furthermore, the same procedures were applied to the connections among the various timber elements of the roof (as well as to the connections between roof and masonry walls, etc.).

The structural analyses show that all wood elements in sound conditions were subject to acceptable stress levels under the prescribed loads, and therefore they fulfilled the safety requirements. On the other hand, the structural checks carried out on damaged wood elements allowed verifying their suitability to work in safe conditions by following a standard approach here synthesized. First, each damage was included as reduction of section into each element or part of it, in order to determine a more correct distribution of the internal forces (axial force, shear and bending moment). It must be pointed out that in this phase the actual wood species and structural grading were taken into account for each element and that the design loads (and not the current ones) were considered. Afterward, the stress level of each damaged element was computed as the "acceptable stress level (ASL)", determined as the "Ultimate Limit State (ULS)" provided in Eurocode 5 [1].

The mechanical characteristics of each wood species were obtained from UNI 11119 [4], which provides strength values in terms of admissible stresses (AS); in order to pass from these AS to equivalent characteristic values, the design value for the ULS (which takes account of both the safety coefficient γ_M and the modification coefficient, k_{mod}) has to be equal to the design value obtainable adopting the AS, given by multiplying the same AS by the suitable load safety coefficient γ_Q^*.

The various typologies of intervention carried out on the elements considered unsuitable based on the described structural check are described in Section 3.6.2, together with the inspiring criteria adopted in the choice of the adopted intervention technique.

Preliminary to every type of intervention is the selection of the wooden material to be used to make repairs.

* The conversion from admissible to characteristic values was carried out according to the criterion briefly described below.

The values reported in UNI 11119 actually refer to a moisture content of 12% and include safety coefficients that take account of the simultaneous presence of both permanent (e.g. own weight) and accidental loads of medium duration (e.g. snow).

On the other hand, in the ULS method, the strength values are characteristic ones, f_k, which have to be modified, introducing suitable coefficients such as the safety coefficient γ_M and the modification coefficient, k_{mod}, which takes into account the service class and the duration of the load. This way the design values, f_d, are obtained by:

$$f_d = f_k \times k_{mod} / \gamma_M$$

Thus, in order to pass from the AS provided in UNI 11119 [4], σ, to equivalent characteristic values, the design value for the ULS has to be equal to the design value obtainable adopting the AS, that is, σ increased by the load safety coefficient, γ_Q:

$$f_d = f_k \times k_{mod} / \gamma_M = \sigma \times \gamma_Q$$

This equation allows determining the equivalent f_k values when suitable k_{mod}, γ_M and γ_Q are used. For k_{mod}, the value related to the average load duration class and the Exposure class 1 is adopted (these are the same conditions at which AS have been obtained); as for γ_M, the value for solid wood is adopted; concerning γ_Q, the value for accidental loads is used. Based on both the Eurocode 5 [1] and the 2008 Italian Code [3], these values are:

$$k_{mod} = 0.8; \gamma_M = 1.3; \gamma_Q = 1.5.$$

Therefore, to pass from AS to characteristic values the conversion coefficient 2.44 to increase σ can be adopted.

3.6.1   Selection of the timber used for the repair interventions

In general, timber used for structural purposes must be strength graded in order to ensure that the wood mechanical performances are in fact those required by design calculations. Grading methods and assignments to strength classes are actually available only for timber belonging to some species and coming from specific geographical areas.

The timber of the roof structural elements of the Nativity church belongs mostly to deciduous oak and, more limitedly, to larch (Lebanon cedar was also occasionally found).

Thus, considering the availability of sellers of old oak elements in Italy (taken from some historic abandoned sites), for which grading rules are available, it was decided to use these elements for repairing the damaged structural elements of the roof and for replacing them in the few cases where it was strictly necessary. The structural quality of each of these elements was checked directly at the sale site by means of the grading rules (specific for oak) provided in the Standard UNI 11119 [4] (as usually occurs for the elements in on-site surveys; Figure 3.6.1). These grading rules allowed assigning a strength and a stiffness class (from I to III) to each analyzed element, which was then marked accordingly. Those elements that were graded as unsuitable for structural use were discarded. A similar approach was also used for the larch elements. When very long oak pieces were needed (in the case of replacements, for instance), it was decided to stock up with new seasoned elements, structurally classified by the seller. Two shipments reached Bethlehem on January 2, 2014, and May 12, 2014.

Actually, the choice of using old wood elements presented two additional advantages: (a) old elements were well seasoned; (b) it was possible to select those elements with a minimal amount of sapwood (although a certain quantity of sapwood was tolerated). Considering that oak heartwood is recognized to be durable against insects (according to EN 350:2016 [5]), a low quantity of the less durable sapwood allowed avoiding the use of any particular systems to preserve wood against possible attacks by both insects and fungi (considering that a suitable aeration system was adopted for the various beam ends, as detailed later). Indeed, the heartwood of both oak and larch has actually proven a natural resistance against insects, as evidenced by the presence of attacks only on the sapwood of the church elements.

The aspect described in point (a) above is particularly important for oak, which is known to exchange moisture very slowly (mainly when mostly made of heartwood). In fact, these old elements, intended to be used for replacing the decayed material, were stored in a sheltered warehouse close to the church (Figure 3.6.1), not in direct contact with the soil, where they were able to adapt to a climate similar to that of the church and to acquire a moisture content similar to the one of the wooden elements inside the church (this difference was always within 5%). That was particularly guaranteed for the cases in which techniques using epoxy resins were adopted for the interventions (see Section 3.6.2 for details) in order to ensure minimal relative movements (due to partial shrinkage/swelling) between the affected parts (the original beam and the prosthesis) just after the gluing operations. Only in the few cases of total replacement of the element was the moisture content not taken into account. Of course, in these cases no adhesive was used.

A similar approach was also adopted to choose the elements for the replacement of damaged purlins: old, well-seasoned boards were used as substitutes of the decayed ones. Heartwood oak boards were mostly used as replacing elements, due to oak's high durability, mechanical strength and stiffness, and high availability (compared to, e.g. Cupressaceae and cedar timber).

Figure 3.6.1 Old elements stored in a sheltered warehouse close to the church.

3.6.2 Intervention typologies

The main guidelines followed in the interventions were the conservation, as far as possible, of the original material and the refusal of visually and aesthetically impacting consolidation techniques, such as those using steel plates or iron girders, which might also raise some problems in case of gluing. In fact, most of the roof structural elements were made of oak. This species is generally characterized by high density, significant and uneven movements in case of moisture variations and presence of localized grain deviation. All of these occurrences imply that the continuous shrinking/swelling movements, which always characterize timber, could not be easily followed by steel and glue, thus causing unpredictable stresses at the edges of the gluing, accompanied by delamination and limited long-term durability.

Moreover, particular care was taken in favoring the exchange of moisture at the truss ends inserted into the wall. In fact, as is well known, fungi may attack wood only when its moisture content is higher than a certain threshold (usually > 20–25%). Thus, facilitating the moisture exchange at timber ends is an obvious way of preventing fungal attack. For the church elements, the following were decided:

- keeping the direct connection between the tie-beam end and the external environment, as is at present;
- keeping the original small niche usually found within the wall above the rafter end (Figure 3.6.2);
- wrapping the sides of both the tie beam and the rafter (including the rafter extrados) within the wall with a 10 mm-thick cork layer in order to avoid both the direct contact between plaster and timber and the moisture absorption from the wall;
- keeping the integrity of the wooden lacings on which the cantilevers are actually resting (Figures 3.6.3, 3.6.4).

All the replaced parts of trusses were catalogued and stored in a suitable place. Shop drawings of the interventions, technical reports and interventions made were checked by members of the Consortium team appointed by the Presidential Committee as supervisors. Interventions could be carried out only upon the approval of the Consortium.

Figure 3.6.2 Original niche in the masonry hosting the cantilever, tie-beam and rafter ends.

Figure 3.6.3 The two wooden lacings (one internal and the other external) on which the cantilevers rest.

Figure 3.6.4 View of the external wooden lacings. The ends of other wood elements (cantilever, tie beam) connected with the external environment are also visible, together with the anchor rod connecting the trusses with the walls through the cantilevers.

The adopted typologies of interventions can be schematically grouped as per below:

1   repair of the damaged beams with timber prostheses connected with glued-in rods (3.6.2.1);
2   repair of damaged beams with pinned half-lap joint (3.6.2.2);
3   repair of damaged beams with self-tapping screws (3.6.2.3);
4   repair of weakened elements with glued-in rods (without any substitution) (3.6.2.4);
5   replacement of totally decayed members with new members (3.6.2.5);
6   replacement of damaged boards and purlins (3.6.2.6).

Before carrying out the interventions on the trusses, it was necessary to remove not-decayed boards and purlins. These removed pieces were catalogued, retained and, after the consolidation, put back in the same place where they were originally. Moreover, some stones of the walls were removed from the support areas of the elements in order to make work easier. They were put back in their original positions when the operations were concluded.

At the end of the interventions on the damaged elements of trusses, and before the complete re-assembling of the roof, the roof structures were tightly connected with the masonry walls in order to improve the seismic response of the church in the case of earthquakes (see Section 3.8).

3.6.2.1 Repair of the damaged beams with timber prostheses connected with glued-in rods

The interventions consisted in repairing the damaged parts of beam-ends (rafters, tie beams or both when actually needed) by using an innovative technique based on the use of prostheses connected to the rest of the original element by steel bars glued with a structural adhesive specific for timber structures, which were placed in slots carved in the wood and then covered by wood laths. Figure 3.6.5 shows the main steps of the intervention procedure in

a)

b)

c)

d)

e)

f)

Figure 3.6.5 The various operating procedures adopted for the interventions, making use of timber prostheses connected with glued-in rods in truss LNN11 of the north aisle.

truss LNN11 of the north aisle, taken as a paradigmatic example. In order to make the inter-ventions as minimally intrusive as possible, the connecting steel elements were confined to the sides of the beam (typically the four corners), and their length reduced to the minimum necessary. In this way, the final aspect of repaired members was excellent and the interven-tion hardly identifiable, so much that it was necessary to affix a small steel plate on each repaired end with the date of the intervention.

The decayed part was cut as close as possible to the sound part of the original beam (identified according to the diagnostic survey), taking into account that glued-in rods had to be glued to sound portions. Great attention was paid to the original geometry of the wood-to-wood joint area so as to reproduce in the new prosthesis the same configuration as the original one. For instance, the same inclination of the contact planes in the joint area was reproduced also in the new elements.

Where and when necessary (e.g. for rafters), purlins were removed starting from the walls up to at least 0.5 m far from the section where the cut was executed in order to facilitate the operations of withdrawal of the decayed part and insertion of the prosthesis (Figure 3.6.5a). Instead, the removal of both the roof external layers (e.g. lead) and the boarding was only necessary in a very limited number of cases. Of course, all elements taken out were put back in the same position after the intervention.

Before carrying out the interventions, the element to be repaired and the related prosthesis were put in contact with one another in order to verify whether they were well aligned along the geometrical axis of the original element (Figure 3.6.5a). Then the two elements were mutu-ally blocked using temporarily screwed small boards (or similar) in order to keep them in place correctly (that is, in their final positions, Figure 3.6.5b). All subsequent working operations (e.g. slots preparation, rod insertion, gluing etc.) were executed on these blocked elements.

The tools used to make the slots were well sharpened and their rotational speed selected so as to avoid any localized wood burning during cutting operations.

The adopted operating procedures can be described as follows:

- propping of the truss (Figure 3.6.5a);
- removal of the decayed end by an inclined cut;
- substitution of the decayed end by the wood prosthesis, shaped in such a way that exter-nal dimensions are the same (Figure 3.6.5a);
- execution of the slots hosting the connecting rods (Figure 3.6.5b);
- partial filling of the slots by the adhesive and introduction of the connecting rods inside the slots (Figure 3.6.5c);
- insertion of wooden laths hiding the rods (Figure 3.6.5d);
- cleaning of the surface and removal of the connection boards;
- removal of the propping after the complete curing of the adhesive (Figure 3.6.5e);
- external treatment of the surface in the intervention area so as to get the same "patina" as the rest of the truss (Figure 3.6.5f).

The final aspect of the intervention for the representative case of truss LNN11 is shown in Figure 3.6.5f. In Figure 3.6.5e and Figure 3.6.5f is also shown the aspect of the repaired beams before and after the final finishing procedure, aimed at making similar the prostheses and the original beams (although they were distinguishable at close sight).

It is important to note that this kind of intervention allows keeping unaltered the original load distribution and constraint conditions of the original truss (e.g. transforming a hinge into a locked system was avoided), thus ensuring on one hand the respect of the construction rules adopted in the past and, on the other hand, the preservation of the original mechanical behavior of the joint (and therefore of the repaired truss).

3.6.2.2   Repair of damaged beams with pinned half-lap joint

This technique consisted in repairing the damaged parts with wood prostheses connected to the rest of the original element using half-lap joints, whose faces were joined to each other using dowels, screws and/or bolts (an example of this kind of interventions is reported in the scheme shown in Figure 3.6.8). Figure 3.6.6 shows some steps for the example of truss LNS08 of the south aisle. The minimum possible amount of connecting pins was used in the interventions, in order to minimize its impact and the waste of original material. Thus, also in this case it was necessary to affix a small plate on each repaired end in order to identify the prostheses.

The general criteria and the adopted operating procedures were similar to those described in Section 3.6.2.1. The dowels were fixed to sound wood, and the cut was made thus to have a straight halved joint (only in a few cases did the joints have oblique faces). Moreover, wooden caps were used to hide dowels and bolts (Figure 3.6.7). As for interventions with glued-in rods, also in the present case the technique allowed keeping unaltered the original load distribution and constraint conditions of the original truss. The same also applies to interventions described in Sections 3.6.2.3 to 3.6.2.5, and hence this occurrence will not be repeated any more.

Figure 3.6.6 The various operating procedures adopted for the repair of damaged beams with pinned half-lap joint for the representative case of truss LNS08 of the south aisle.

Figure 3.6.7 Examples of wooden caps and plates used to hide dowels and bolts.

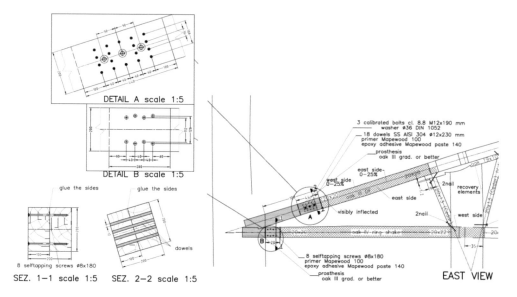

Figure 3.6.8 Scheme of the intervention making use of pinned half-lap joint carried out on truss LNS08 of the south aisle.

The present technique was preferred to one making use of glued-in rods in the following two cases:

a   When it was needed using new and not completely seasoned timber for the prostheses, due to the lack of old elements whose section was as large as the original elements;
b   When the prosthesis to be used was very short, thus making unreliable the connection with rods that were not long enough.

3.6.2.3 Repair of damaged beams with fully threaded self-tapping screws

The technique was adopted when it was necessary to perform local consolidations to reinforce elements with deep and/or extensive cracks judged as possibly causing crack propagation

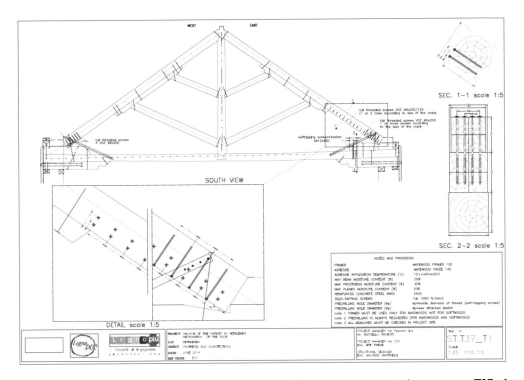

Figure 3.6.9 Scheme of the intervention making use of self-tapping screws carried out on truss T37 of the south transept.

during time based on the diagnostic survey. Interventions consisted in using self-tapping screws as sewing elements between undamaged parts of the same timber member. Screws were positioned so as to respect the geometrical distances prescribed in applicable codes. In oak wood pre-drilled holes were executed in order to limit as much as possible the potential stresses induced by screwing operations, also considering that the interest areas were partially damaged. Moreover, headless screws were used, thus allowing their complete insertion into the element to be reinforced. After screws' insertion, holes were filled with either wood stoppers or putty. An example of this kind of interventions is reported in the scheme shown in Figure 3.6.9.

3.6.2.4 Repair of weakened elements with glued-in rods (without any substitution)

In a very few cases (the tie beam of truss CN12 and the king-post of truss TN17) the elements were in a potentially unsafe condition but, nevertheless, not so worrying as to require their partial or total replacement. In the tie beam CN12, for instance, a very big knot in the tie beam had reduced considerably the dimensions of the transversal resisting section. As is obvious, in this case the reduction in strength was due to a natural defect existing in the element and not to the applied load conditions. It is worth noting that a thin nailed iron belt (Figure 3.6.10)

Figure 3.6.10 Aspect of the tie beam CN12 before making interventions. A big knot has been reinforced in the past using an iron belt wrapping the element.

Figure 3.6.11 The transversal crack weakening the king-post of truss TN17.

wrapping the tie beam was already present (almost surely applied when the tie beam was put
in place), but it was considered not efficient according to the structural analysis.

In the case of the king-post of truss TN17 in the north transept there was a transversal
crack due to a partial rupture of the element (Figure 3.6.11), which probably occurred during
the bombing of this transept, that caused also the collapse of the nearby truss (the rafters of
which are in fact made of new pine elements instead of oak).

In both these cases, the structural analysis showed that the stress levels in these elements were limited (or even very limited) and that the residual sections could be considered as sufficiently resistant. Thus, it was decided to avoid the unnecessary substitution of any portion of the original elements while reinforcing them so as to ensure an adequate level of structural safety. In both these cases, the interventions were carried out as in the case of repairs by means of glued-in rods (Section 3.6.2.1), with bars "sewing" the weakened part of the element (Figure 3.6.12 and Figure 3.6.13).

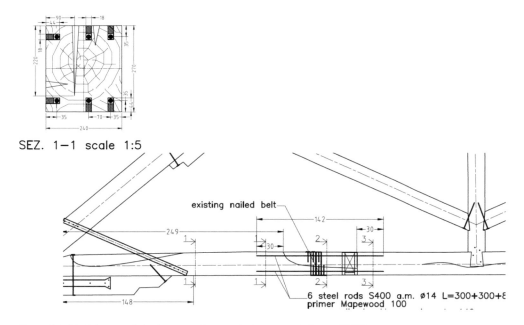

SEZ. 1—1 scale 1:5

existing nailed belt

6 steel rods S400 a.m. Ø14 L=300+300+8
primer Mapewood 100

Figure 3.6.12 Scheme of the intervention carried out on tie beam CN12.

Figure 3.6.13 Aspect of the king post TN17 after the reinforcing intervention with glued-in rods carried out without using any additional prosthesis.

It is also worth mentioning that the iron-nailed belt of tie beam CN12 has been re-put in place in order to keep memory of the original intervention.

3.6.2.5 Replacement of totally decayed members with new members

Based on the survey described in Section 3.4, only selected elements needed to be totally replaced with new members. The final decision of substitutions was also based on numerical investigations taking into account both the current load conditions and the presence of the damage. Thus, the replacement was taken into account only for the elements not capable to bear the considered loads.

This kind of intervention concerned cantilevers (Figure 3.6.14 a) and entire rafters (Figure 3.6.14 b).

Before carrying out these interventions, geometrical measurements (base, height and possible wanes) were taken as closer as possible to the nearby elements, thus to reproduce in the new timber element the same configurations as the original ones.

On the other hand, it was very difficult to preserve any single original nail (e.g. connecting rafter and tie beam, including those used for the external iron passive bands, or connecting tie beam and cantilever) in its position. In fact, due to either the oak strength or non-linearity of the nail stem, several of these nails broke during the removal process, although it was carried out very gently. In these cases, new screws were placed to secure the joints, close to the original positions of nails, without altering the authentic mechanical scheme of wood-to-wood connections. Before inserting these long screws, the affected element was pre-drilled, in order to prevent breaking the screw owing to excessive torsion. Of course, the pre-drill diameter was lower than the screw section.

In a similar way, the original nails between purlins and the rafter to be substituted were recovered each time it was practically possible. In fact, only a few screws were used to substitute the broken nails in purlins.

When the intervention affected the cantilevers, it was ensured that they were also connected to stringcourses, as it was for the original ones.

The highest percentage of replaced elements was in the crossing area, at the intersection of the nave with the transept, where both rafters of one diagonal truss were totally replaced,

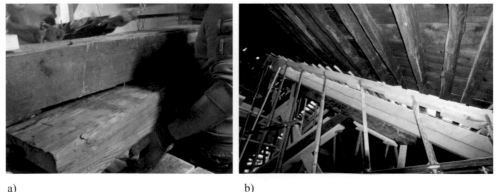

a)   b)

Figure 3.6.14 Replacement of a cantilever in CN 04 (a) and a rafter in CN 07 (b).

Table 3.6.1 Existing timber elements – Replaced timber elements

Area	Total length of existing timber elements (m)	Total length of replaced timber elements (m)	Percentage of replaced timber elements (%)
Central nave	604,71	54,47	9,01
North transept	240,22	21,73	9,05
South transept	240,22	12,79	5,32
Apse	240,22	10,66	4,44
Crossing area	115,43	28,15	**24,39**
North corner	95,24	9,35	9,82
South corner	95,24	4,81	5,05
North aisle	327,96	27,79	8,47
South aisle	327,96	10,02	3,06
	Total 2287,20 m	**Total 179,77 m**	**Total 7,86%**

as shown in Table 3.6.1. However, it is worth noting that less than 8% of all existing truss elements were replaced as shown in Table 3.6.1.

3.6.2.6 Replacement of damaged boards and purlins

The survey of both boards and purlins was carried out at both the intrados (visible from the inner of the church) and the extrados (not visible) of the elements, of course in this latter case after the removal of the roof covering. The decision about replacing or keeping in place the elements was carried out by adopting the scheme shown in Figure 3.6.15 under the control of qualified personnel, highly expert in the evaluation of wood decay.

The adopted general criteria were the following ones:

- the boards were assessed both visually (darkening possibly due to brown rot and whitening possibly caused by white rot) and by means of penetrometric tests (use of blunt-pointed tools able to gently probing the surfaces to reveal the presence of softened areas). Boards were considered severely attacked when the testing tool could easily penetrate inside them and when the decay concerned at least 50% of the board surface. This was considered suitable, from a structural perspective, to allow for an efficient cooperation between the boards and the plywood layer intended to be put above them;
- considering the huge number of purlins present in the roof covering and their very close spacing, the purlins were only assessed for decay when the amount of decayed boards per pitch sector was higher than 50%, because it was considered that in this case an important water leakage affected the considered pitch sector. A pitch sector is half the area of a pitch included in a truss span, and contains on average 12 rows of boards covering the whole pitch, from the ridge to the gutter (Figure 3.6.16).

On the other hand, particular attention was paid to the purlin conditions at the connections with the tympanum wall, because in this case a reliable connection with the seismically active plywood/boarding layers had to be ensured. In that case, the boards were always removed (irrespective of the presence of any biological damage in the same boards) in order

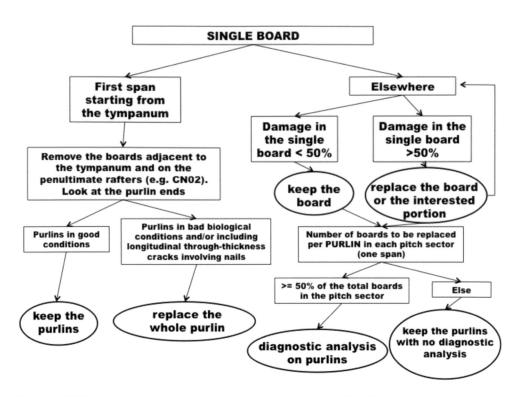

Figure 3.6.15 Decision scheme adopted for replacing or keeping in place boards and purlins.

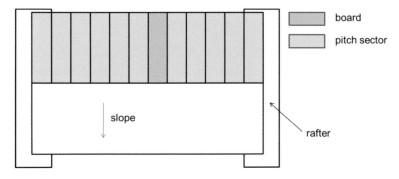

Figure 3.6.16 Scheme of a pitch sector.

to check the conditions of the purlin–rafter connections. Of course, when an extensive bio-logical decay (mostly due to fungal attack, Figure 3.6.17) was present the affected element was replaced, because it was not able to safely transmit any load. Moreover, in the presence of through-thickness cracks (Figure 3.6.18), purlins were replaced because this occurrence prevented them from withstanding loads along their longitudinal axis. In contrast, when the shrinkage crack stopped to the pith (Figure 3.6.19) purlins did not need to be replaced.

The survey carried out on the boards and purlins of the roof evidenced a generally accept-able state of conservation of these elements (Figure 3.6.20): less than 9% of the boards were

Figure 3.6.17 Example of a purlin damaged due to fungal attack.

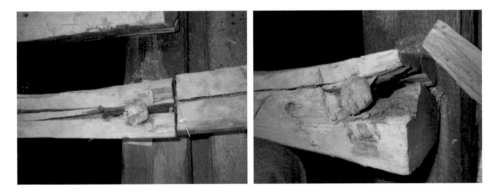

Figure 3.6.18 Examples of purlins with not-acceptable connections with rafters.

Figure 3.6.19 Example of a purlin with acceptable connections with rafters.

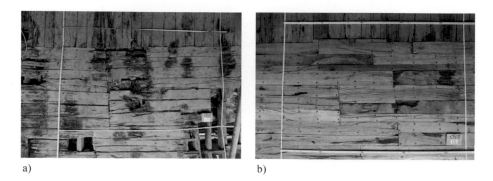

a)   b)

Figure 3.6.20 Examples of partially damaged (a) and of replaced boards (b).

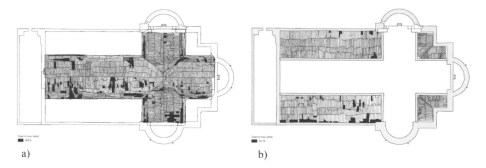

a)   b)

Figure 3.6.21 Map showing the boards replaced in the nave and transept (a) and in the aisles (b).

replaced in the nave and transept (Figure 3.6.21a) and only 6% in aisles and corners (Figure 3.6.21b). Insignificant was the number of replaced purlins. Even these interventions on purlins and boards were made by using old wood brought from Italy.

3.6.3 Structural adhesive used for the repair interventions

The structural adhesive used for the interventions was a bi-component and thixotropic epoxy resin specifically formulated for structural interventions on timber elements and declared able to withstand structural loads safely for a minimum thickness of 2 mm. The manufactures gave evidence of the suitability of the used product as a structural wood adhesive, as this was expressly provided in clause 11.7.7.2 ("Adhesives for joints made on-site") of the Italian Standards [3]. In fact, this law states that adhesives intended to be used for joints made on-site must be subjected to tests according to suitable protocols aimed at evidencing that the joint shear strength is not lower than the one of solid wood tested in the same conditions.

Thus, before choosing the product the manufacturer was requested to bring evidence that the adhesive fulfilled the following requirements:

- it was tested also on oak, in addition to other possible wood species;
- the maximum bonding pressure was 3 bars (a maximum, instead of minimum, value was specified because adhesives to be used on-site are usually applied without any bonding pressure, except for clamping);

- the joint shear strength had to be not lower than that one of solid wood tested in the same conditions.

The manufacturer gave evidence that the adhesive was tested in both dry and wet conditions, in this latter case after the application of preliminary treatment cycles. They consisted in a repeated immersion in water under pressure and under vacuum and then to a rapid drying before testing in shear. Similar conditions are also prescribed in EN 302–2:2004 [6] or EN 14080:2013 [7]. As an example, in those conditions solid spruce wood evidences a shear strength value of 8 N/mm^2 when tested dry and 4 N/mm^2 when tested wet. According to the adhesives manufacturer, the requested requirements were fulfilled by using a primer (also of epoxy type) applied to the wood surface prior to bonding.

References

[1] EN 1995–1–1:2008 (Eurocode 5) *Design of timber structures*. Part. 1–1: General-Common rules and rules for buildings.
[2] EN 1998:2005 (Eurocode 8) *Design of structures for earthquake resistance*. Part 1: General rules, seismic actions and rules for building, Part 3: Assessment and retrofitting of buildings.
[3] Norme tecniche per le costruzioni – D.M. 14 Gennaio 2008.
[4] UNI 11119:2004 *Beni culturali – Manufatti lignei – Strutture portanti degli edifici – Ispezione in situ per la diagnosi degli elementi in opera* (Wooden artefacts: Load-bearing structures: on-site inspections for the diagnosis of timber members).
[5] EN 350:2016 *Durability of wood and wood-based products: Testing and classification of the durability to biological agents of wood and wood-based materials.*
[6] EN 302–2:2004 *Adhesives for load-bearing timber structures*. Test methods, Part 2: Determination of resistance to delamination.
[7] EN 14080:2013 *Timber structures: Glued laminated timber and glued solid timber: Requirements.*

3.7 Archaeological analysis

G. A. Fichera

The archaeological analysis of the architecture of the Nativity church, carried out during the first diagnostic campaign in 2009–2010, used the analysis tools typical of the archaeology of architecture, an investigation methodology never applied in the past to the study of the basilica. The study focused in particular on the walls of the church and on the relationship between these and the complex system of caves below, omitting the analytical analysis of the fortified complex built around the basilica from the end of the 11th century, date of conquest by the Crusaders.

The most important research carried out during the 20th century (see the Introduction to the book) had favoured a type of investigation based above all on the documentary sources and on analysis criteria borrowed from the history of art or architecture, arriving at interpretations also quite discordant in relation to the real era of construction of the building and to the main transformations this has undergone over the centuries.

The stratigraphic analysis of the architectures has made it possible to clarify the constructive sequences of the monumental complex thanks to objective analysis criteria and extensively tested in other contexts of study that have allowed them to identify all the main constructive and destructive actions that have marked the life of the building over time, identifiable thanks to the recognition of the different masonry stratigraphic units and thanks to an extremely analytical sampling of the construction techniques used, an important fossil guide in the reconstruction of the building phases of the basilica itself.

The main results of the study, already published in international journals and conference proceedings [1][2], attest to a constructive sequence that shows a relationship of substantial contemporaneity between the narthex, the naves and the tri-apsed system located to the east. Therefore, once established that all the architectures go back to a precise historical moment, thanks to the archaeometric analyses carried out on the architraved system that dominates the columns, it is now possible to obtain a certain and absolute dating of that historical moment.

The analyses carried out on the wood species that make up the architraves, which will be discussed in detail in Chapter 8, confirm that the construction of the basilica dates back to the mid-sixth century AD. C. The architraves are part of a unitary construction system linked to the construction of the overlying walls, in turn linked to the perimeter walls. Therefore, it is absurd to think that they could have been replaced in the past, when some delicate intervention techniques, we could use nowadays, were absolutely not available. The Nativity church is, therefore, the result of a major construction project undertaken by Justinian and perhaps completed in the decades following his reign. This project, in addition to the construction of a new impressive basilica, made accessible and expanded the caves below, with new and monumental entrances, with the aim of creating new and more detailed circuits for pilgrims.

The stratigraphic analysis, supported also by the one referred to the construction techniques, was able to confirm the contemporaneity of the entire building and demonstrated that the two constructive joints or brake lines placed in a perfectly symmetrical position at the end of the two transepts, in the past hypothesized as signs or traces related to different construction periods, were actually due to breaks in the same construction process belonging to the same chronological context (Figures 3.7.1, 3.7.2), in analogy with what was done in buildings in Central and Northern Italy [1].

Figure 3.7.1 Western wall of the north transept.

Figure 3.7.2 Western wall of the south transept.

The presence of these break lines and of other stratigraphic relationships made it possible to determine the sequence of the construction phases of the basilica. In particular, the construction of the lower part of the church, including the narthex, started from two opposite sides, the narthex to the west and the three apses to the east, probably to allow the use of the presbytery for religious functions and a protection to the grotto in the shortest times. These two bodies found a junction point at the intersection of the transept with the nave, just where the previously mentioned joints are now visible, immediately below the windowsills.

From that level the construction of the upper part of the walls proceeded simultaneously around the whole perimeter of the church, including the 34 arches necessary for the construction of the windows [1].

The construction technique surveyed in all the walls attributed to the original construction phase is characterized by the use of perfectly squared blocks of large dimensions (up to 1 m in length and up to 0.4 m in height) placed on horizontal and parallel rows of fairly regular but not always identical heights. All mortar joints are extremely thin thanks to the perfectly flat and adherent faces of the blocks (Figure 3.7.3).

Although over the centuries cement mortars were used to cover the original joints, the recent analytical reading made it possible to identify the traces of the original mortar, very compact whitish lime mortar, rich in shredded brick fragments, also covering the edges of the individual blocks and sometimes engraved with a pointed tool (Figure 3.7.4).

The traces of the working tools refer to a flat-blade chisel used to make the contour of the stone block (anathyrosis), 2/3 cm thick, and to two different types of chisel (toothed-blade chisel) used for the finishing of the block surfaces (Figure 3.7.4).

The construction technique is associated with the local stone lithotype, the so-called malaki, with chromatic nuances from light yellow to grey and with different levels of compactness.

Figure 3.7.3 Regular and thin mortar joints.

Figure 3.7.4 Finishes at the stone edges.

Between the end of 2013 and 2014 the restoration works on the roof covering allowed us to observe in detail also the top portions of the walls placed in direct contact with the wooden trusses. In particular, along the nave and transept, immediately under the cantilevers of the trusses, it was possible to survey the presence of a continuous system of double wooden lacing on both the inner and outer side of the walls. Above these wooden lacings the perimeter walls continue for other two rows of stone that support the series of purlins forming the roof eave (Figure 3.7.5).

However, the diagnostic analyses showed that both the wooden lacings and the last two rows of stone were replaced after the construction of the basilica during one of the main interventions of restoration of the roof. The non-belonging of such top portions of wall to the same construction period of the whole church is archaeologically proven by a different construction technique, by the re-use of stones taken elsewhere and by the use of a macroscopically different mortar. Even the wooden species of the double lacing is different from that characterizing the original wooden parts of the church (see Section 3.5). In detail, the stones used at this level, although they are the same type used in the rest of the church, have different dimensions, and the mortar used is not characterized by the presence of brick fragments ("*coccio pesto*"), which are instead macroscopically evident in the original mortar. Also the stratigraphic relationships between the different wall blocks confirm the re-construction of the upper level in later time, as evidenced at the intersection between the external wall of the south aisle and the counter-façade of the church, where the masonry was broken to host the external wooden lacing (Figure 3.7.6).

Figure 3.7.5 Rows of stone added above the double wooden lacing.

Figure 3.7.6 Counter-façade wall broken by the external wooden lacing in the south aisle.

Archaeological traces of the interventions of reconstruction of the roof, which occurred over the centuries, were brought to light also in the aisles where the removal of the roof brought out a series of architectural elements related to a roofing system set to a slightly higher level than the current one and which can be traced back to Justinian's era, mainly on the basis of the peculiar processing of the lithic elements.

In particular, they are the cavities placed in the external walls of the north and south aisles to host the truss ends (Figure 3.7.7) and the circular holes (diameter 15 cm) in the western side of the transept and in the eastern side of the counter-façade, placed at a

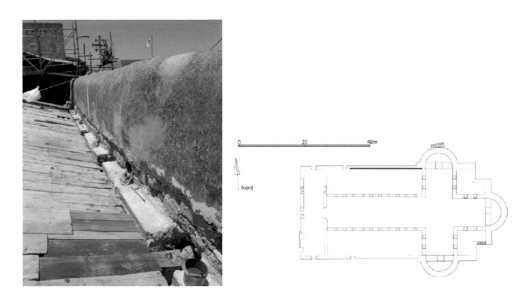

Figure 3.7.7 Cavities in the external wall of the north aisle.

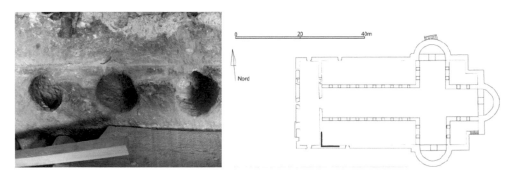

Figure 3.7.8 Circular holes for original purlins.

constant distance of 14 cm from each other (Figure 3.7.8). This series of holes, perfectly carved in the stones of the original masonry, had to host the ends of the purlins placed above the wooden trusses.

Moreover, in the walls of the nave, just below the windowsills, a linear recess was found, used in the past to host the upper part of the pitch (Figures 3.7.9, 3.7.10).

Further archaeological proof of the fact that the original roofing of the aisles and corners was set at a higher level than the current one is attested by the discovery of a recess, in correspondence of the north-eastern corner of the basilica, with a quadrangular cross section, hosting the internal wooden lacing that supported the original truss end. This lacing was removed during the reconstruction of the roof, but some wood traces are still preserved on the surface of the mortar.

Figure 3.7.9 Linear recess at the higher level of the aisle pitches.

Figure 3.7.10 Linear recess at the higher level of the aisle pitches.

The new truss end is now placed at a lower level, as shown in Figure 3.7.11, directly upon the key stone of the arch of one of the windows of the lower level (Figure 3.7.12). It is worth noting that the integrity of the external wall surface shows that there has always been only one wooden lacing placed on the inner side of the wall, unlike what occurs at the upper level of nave and transept, where this lacing is on both the inside and the outside.

Figure 3.7.11 Recess in the wall of the north-east corner.

Figure 3.7.12 Keystone corresponding to the truss end.

References

[1] Bacci, M., Bianchi, G., Campana, S. & Fichera, G. (2012) Historical and archaeological analysis of the Church of the Nativity. *Journal of Cultural Heritage*, 13, Elsevier, e5–e26.
[2] Fichera, G., Bianchi, G. & Campana, S. (2016) Archeologia dell'Architettura nella Basilica della Natività a Betlemme, pp. 1567–1589, in Olof Brandt – Vincenzo Fiocchi Nicolai (a cura di) *Acta XVI Congressus Internationalis Archaelogiae Christianae Romae* (22–28.9.2013). *Costantino e i Costantinidi. L'innovazione costantiniana, le sue radici e i suoi sviluppi*. Pontificio Istituto di Archeologia Cristiana, Città del Vaticano.

3.8   Vulnerability analysis and seismic reinforcement

C. Alessandri, V. Mallardo and M. Martinelli

Bethlehem is in a seismic area, near the fault line dividing the African plate from the Euro-Asian plate (Figure 3.8.1) and with a peak ground acceleration between 0.15 g and 0.20 g (Figure 3.8.2). More or less severe earthquakes have always struck the Middle East area, in particular Palestine and Syria. The most ancient historical testimonies are provided by letters written by monks and priests in Latin, Greek and Arabic. The earliest earthquakes mentioned in these testimonies date back to the first decades before the birth of Christ and follow one another in successive epochs at different intervals and with varying intensity. The

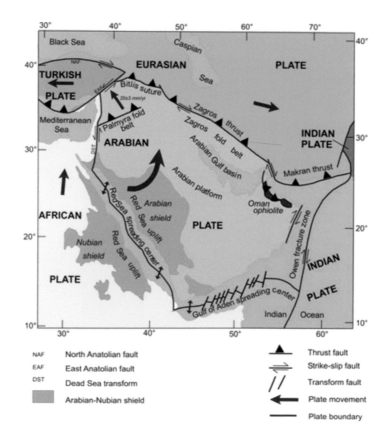

Figure 3.8.1 Fault lines.

Seismic Zone Factor,Z

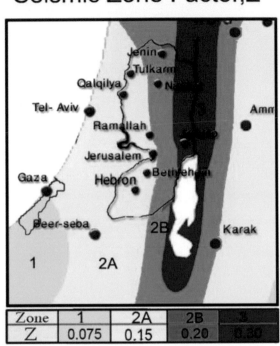

Zone	1	2A	2B	3
Z	0.075	0.15	0.20	0.30

Figure 3.8.2 Seismic zones.

most devastating in terms of injuries and deaths were those of 1457, 1752 and 1837–38, also accompanied by considerable damage to civil and religious structures.

Nowadays it is still rather difficult to bring ancient buildings up to the modern safety standards and to preserve, at the same time, their historical, architectural and artistic features. On the other hand, the concept of "seismic improvement" has been spreading more and more among the structural engineers involved in conservation and restoration and is now clearly stated in the current technical standards: rather than increasing the seismic performances up to the level required for new buildings, it is advisable and more reasonable to improve them as much as possible without making any significant change to the main characteristics of the building. This is the main criterion that was followed in the interventions, basically aimed at improving the connections between the restored roof and masonry walls so as to transfer part of the horizontal seismic action to the shear-resistant walls of the church.

The structure already has good if not excellent anti-seismic characteristics. In fact, the ancient designers thought up a very robust structure, already equipped with many of those devices that nowadays are used to improve the seismic characteristics of buildings.

The trusses of the nave are widely dimensioned, with rafters that, for their span and slope, have large cross sections, equally large under-rafters that cover almost all the length of the rafters, double struts, very large and long sleepers and king posts fixed with nails to the tie beam (unlike what happens in the common Palladian trusses). It is worth remembering that all the roof components are already tightly connected to each other with iron nails of different sizes (Figures 3.8.3, 3.8.4), e.g. wooden boards with purlins, purlins with rafters, rafters

Figure 3.8.3 Nails between rafter and tie beam.

Figure 3.8.4 Nail crossing the truss end.

with tie beams and sleepers at both ends of each truss, sleepers with wooden lacing, so that the whole roof is like a rigid hat resting on the perimeter of the church.

Moreover, each sleeper is provided with a horizontal iron rod nailed to one side and anchored to the external surface of the wall through a vertical iron rod. This is a fairly common expedient in the anti-seismic adaptations of buildings to avoid the sliding of the floors off the walls and to limit the relative horizontal displacements between two opposite walls (Figures 3.8.5, 3.8.6).

A double wooden lacing (Figure 3.8.7) is placed along the whole perimeter of the upper part of the church (nave, transept and apse), immediately under the trusses and usually fixed to thcm and to the central core of the walls by means of five large forged nails. In particular, (Figure 3.8.8), the nails are $16 \times 16 \times 360/380$ mm with round head (diameter 45 mm), four of them are driven into the double lacing, two on each side and one into the masonry that fills the inner space between the two lacings (Figure 3.8.9).

Figure 3.8.5 Nailed horizontal iron rod.

Figure 3.8.6 Vertical iron rod.

Figure 3.8.7 Double wooden lacing.

Figure 3.8.8 Nails connecting truss end with lacings.

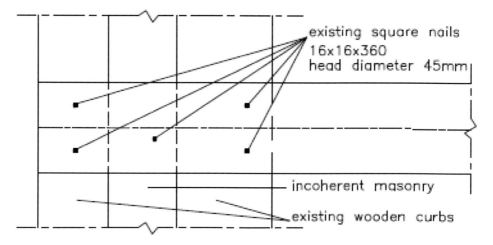

existing square nails
16x16x360
head diameter 45mm

incoherent masonry

existing wooden curbs

Figure 3.8.9 Position of nails in the wall thickness.

An inner wooden lacing, hidden by the plaster, was found also in the walls of the aisles, although interrupted in correspondence of the narthex and in the northern wall between LNN6 and LNN8 trusses, where they were removed during some interventions of consolidation carried out in the past. As described in more detail in Section 3.2.1, these wooden lacings are made of consecutive elements joined directly by means of a nailed half-wood head joint (Figure 3.8.10) or by means of an iron rod nailed to one side (Figure 3.8.11). Undoubtedly they contribute to increase the rigidity of the whole roof and to improve its connection with the masonry structure underneath.

Figure 3.8.10 Nailed head joint.

Figure 3.8.11 Head joint with nailed iron rod.

Other reinforcements, such as external small pillars (Figure 3.8.12) acting as buttresses and iron belts around the protruding elements of the roof structure (Figure 3.8.13), are already present at the corners of the central area where the diagonal trusses intersect the trusses of nave, transept and apse (see Section 3.2.3).

Moreover, at the intersection of orthogonal walls it is easy to find L-shaped angular stones that contribute to improve the connections among these walls (Figure 3.8.14).

Nevertheless, despite such precautions, the church probably suffered some damage due to seismic events in the past, some of which reported in the so called *firmans* (written permissions, granted by Islamic officials, to carry out interventions of restoration in the church).

Figure 3.8.12 Stone reinforcement at the corners.

Figure 3.8.13 Iron belts around timber elements.

The presence of an external buttress on the north side (Figure 3.8.15) and the closure of some windows at the intersection of nave, transept and apse (Figures 3.8.12, 3.8.16) are in fact interventions carried out to remedy some damages suffered or to prevent future damages.

The basic idea of the whole anti-seismic consolidation project was therefore to increase the stiffness of the roof by replacing the old roof covering with a new one in several layers (including plywood) closely connected to the existing planks and purlins by galvanized steel screws. Moreover, the connection between roof and walls was improved by connecting all the truss ends, and in some cases the ends of rafters and purlins, to the underlying walls, as will be explained in detail in the following Section. In this way, thanks to this strong connection among all components of the roof and between roof and walls, all the church can behave as a rigid box with all benefits that such a behavior can provide. The masonry structures of the church

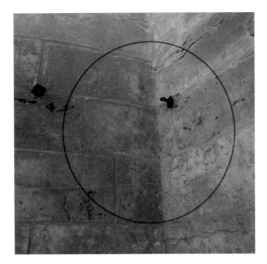

Figure 3.8.14 L-shaped angular stone.

Figure 3.8.15 Buttress on the north side.

Figure 3.8.16 Buffered windows and corner stone reinforcement.

did not show any significant sign of weakness and material decay; therefore, no interventions were planned and carried out on them. Their resistance to the seismic loads was checked in the numerical analyses carried out with reference to the current Italian Technical Standards [1]. These analyses mainly concerned the interactions between the roof structures and the walls in the presence of the added reinforcements, local stress conditions and the behavior of single components of the roof covering under in-plane seismic actions.

3.8.1   Interventions of seismic improvements

The connection between the trusses of the nave with the external walls (at the upper level) was obtained by means of M20 stainless vertical threaded bars, crossing the rafters (wherever possible), tie beams and sleepers up to the masonry where they penetrate for a depth of 30 cm so as to avoid superficial weaker zones and guarantee at least a 15 cm grip inside the good stones (Figures 3.8.17, 3.8.18, 3.8.19). In the aisles the trusses are connected to

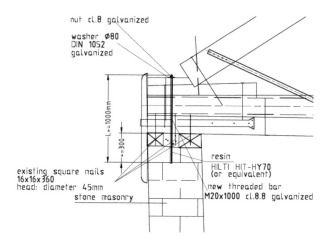

Figure 3.8.17 Vertical bar as seismic connector.

Figure 3.8.18 Vertical seismic connector at the truss end.

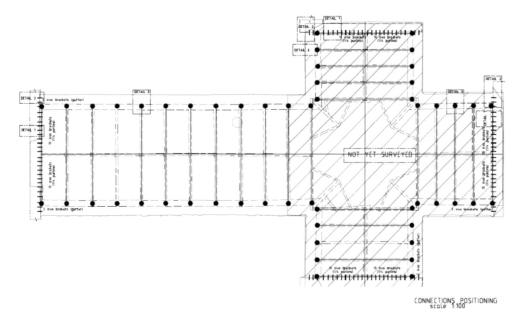

CONNECTIONS POSITIONING
scale 1:100

Figure 3.8.19 Plan view of the seismic connectors (dots) at the upper level.

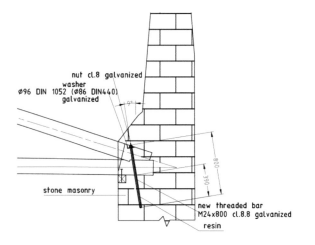

Figure 3.8.20 Inclined bar as seismic connector.

the north and south walls by inclined M24 stainless threaded bars inserted into the masonry for nearly 40 cm (Figures 3.8.20, 3.8.21, 3.8.22). The same bars were used to connect the half trusses of the two corners to the external walls (Figure 3.8.22). These types of bars, as well as other devices used to connect the roof to the walls, are called seismic connectors from now on. In both cases mentioned above a perforation of the truss wooden elements and stones was done before the insertion of each bar. A thin layer of epoxy resin fills the space between stones and steel bars which, on the contrary, are in direct contact with the wooden

Figure 3.8.21 Inclined seismic connector at the truss end.

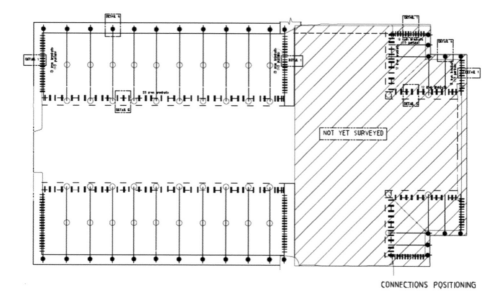

Figure 3.8.22 Plan view of the inclined connectors (dots) at the lower level.

truss elements in the upper part of the connection. In this way a more effective joint is guaranteed between walls and trusses, the components of which might be removed, if necessary, by simply cutting the steel bars at the interface wood-masonry. The position of such connectors (black dots) in the trusses of nave, transept and apse and in the trusses of the aisles and corners is shown in Figures 3.8.19 and 3.8.22, respectively.

Another type of connector was used wherever the use of the connectors previously described, in particular the inclined bars, would have been more difficult for the peculiar morphologic and technical characteristics of the intervention area or for the presence of mosaics on the opposite side of the wall. They are galvanized steel devices, made of a bent rebar (diameter 12 mm, slope and length changing according to the different situations) welded to a rectangular plate (thickness 5 mm, variable width and length) and inserted into the masonry for an approximate length of 300 mm (Figures 3.8.23, 3.8.24). The connection between rebar and masonry is improved by an epoxy resin filling the space at the interface (the same used for the connectors of the trusses).

Figure 3.8.23 Rebar–plate connector.

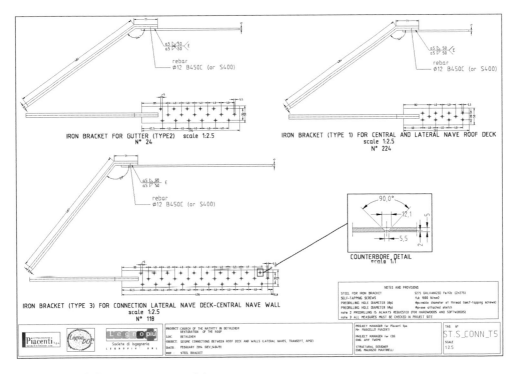

Figure 3.8.24 Different geometries of the same connector.

The connection of the roof of nave, transept and apse with the four tympana (Figure 3.8.19) is obtained by fixing the plate to the purlins and the existing boards by means of self-tapping screws (diameter 5 mm, length 100 mm) inserted immediately under the new plywood (Figure 3.8.25). This type of connection occurs every four purlins, for a total of 10 connections on each pitch. It is a non-invasive intervention, requiring only some holes drilled into the masonry walls, and therefore very common in seismic reinforcement of masonry buildings. The same devices and the same way of application were used to connect the purlins of the aisles with the façade of the church and the west walls of the transept, as well as the purlins of north and south corners (between apse and transept) with the external walls (Figure 3.8.22).

Similar devices with different dimensions were used to connect the roof of the aisles with the walls of the nave and the roof of the corners with the walls of transept and apse (Figure 3.8.22). In this case the plate is a bit longer in order to catch at least three purlins and the boards above them (Figure 3.8.26). At least three connectors are used within each span. For

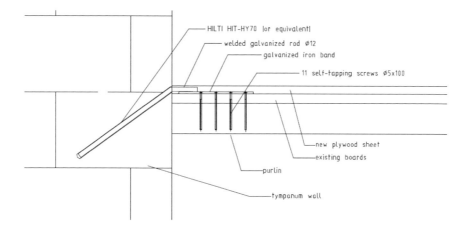

Figure 3.8.25 Connection between wall and orthogonal purlins.

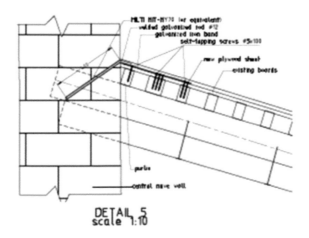

Figure 3.8.26 Connection between wall and parallel purlins.

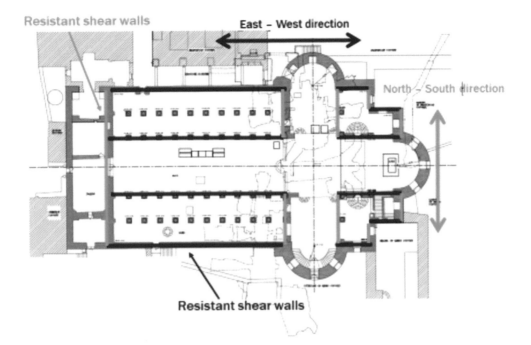

Figure 3.8.27 Resistant shear walls in N-S and E-W directions.

sure, these rebar–plate devices guarantee a more widespread and less invasive link wherever they were used and, together with the threaded bars in the truss ends, firmly anchor the roof to the walls, thus increasing the stiffness of the entire building and its resistance against seismic actions.

Thanks to these connections the horizontal seismic loads acting in the east–west direction (longitudinal direction) at the lower level of the roof can be absorbed completely by the external walls of aisles and corners, whereas only 48% of the horizontal seismic loads acting in the same direction at the upper level of the roof can be transferred to the walls of nave and apse (Figure 3.8.27). In the north–south direction (transversal direction) only the walls of narthex, transept and corners can withstand the seismic actions at the upper level of the roof. The maximum horizontal load, as average value, that can be transferred from the roof to these walls is approximately 42% of the total seismic action computed in this direction.

The interventions carried out were based on the results obtained from two structural analyses referring to different structural models with different levels of accuracy and different correspondence with the actual structural behavior of the church under seismic actions. The first one basically assumes the shear-resistant walls as simple supports for the planks of the roof under horizontal in-plane seismic actions in both east–west and north–south directions. This model is not necessarily in favor of safety, since it assumes the maximum possible seismic action in the hypothesis that the design acceleration falls on the plateau of the spectrum, whereas the torsional effects, either those due to possible eccentricity between the center of mass and center of stiffness or those required by law as a minimum, are neglected.

The latter is a three-dimensional Finite Element model, implemented in the PRO_SAP Software [3], with shell elements for walls and vaults, beam elements for beams and columns and membrane elements for planks. For the sake of simplicity, a linear elastic behavior was assumed for all the materials used. The analyses carried out, including a modal dynamic analysis, provide accurate values of the seismic loads to be applied to the roof structural components and in particular the maximum shear force on each shear-resistant wall.

A particular attention was given to the modeling of the plank and to the evaluation of its effects on the whole structural behavior of the church. The plank is made of relatively thin plywood (21 mm thick), and it could not be assumed as an infinitely rigid plane. Therefore, it was given the same stiffness as the plywood.

The seismic action was evaluated according to Eurocode 8 [2], which considers, differently from the Italian Standards, seismic zones characterized by different reference acceleration values (like in the Palestinian standards). The ground acceleration for Zone 3 ($a_g = 0.15$ g, the same value prescribed for Bethlehem) was considered. In order to obtain a design elastic spectrum with the same peak acceleration value ($a_g = 0.325$) used in the preliminary calculations, an importance factor 1.7 and a site factor $S = 1.142$ were used. The maximum seismic action is foreseen with a return period of 1,898 years. The fundamental period estimated approximately by the pseudo dynamic analysis is equal to 0.426 s; the value calculated by the dynamic analysis is 0.44 s (very close to the approximate value and in any case within the plateau). The seismic calculations take into account the geometrical eccentricities and the additional ones prescribed by the regulations. The seismic combinations are made according to [1].

The modal analysis highlighted that the seismic actions at the lower levels of the roof (aisles and corners) are lower than those estimated by the first approximate approach. In particular, they are differently distributed among the walls according to their stiffness, which confirms that the plank, although assumed with limited stiffness, produces effects very similar to those of an infinitely rigid plank.

The results of the dynamic analysis, apart from small variations, confirm those of the preliminary calculations and the above-mentioned percentages of seismic action transferred to the walls at both levels.

The same 3D model was used to make some comparison among different types of bracing.

By using the same 3D model and the same dynamic analysis, the effects of the structural plywood panels, placed above the existing timber boards and screwed to the secondary roof structure (purlins), were compared with those of a bracing system made of steel crossed bands nailed to the existing timber boards. It is worth noting that this bracing system could have never been used because of the double slope of the roof pitches and the consequent impossibility to bend the bar or stiffen only part of the pitches. However, the comparison highlighted the structural advantage of plywood, as a bracing system. As a matter of fact, for being extremely stiff in its own plane and for being fixed to the underlying elements with a great number of screws widely diffused all over the surface, it allows for a better and more uniform distribution of horizontal actions among the roof trusses. This cannot occur, if not partially, with crossed bars fixed along definite directions and in a more limited number of points.

For both bracing systems the deformed configurations of the church were obtained in the presence of seismic actions in both N–S (X direction) and E–W directions (Y direction). As the values of the horizontal displacements in the X direction are higher and more significant than those in the Y, only the deformed configurations produced by seismic actions in the N–S direction are reported in Figure 3.8.28 for the plywood system and in Figure 3.8.29

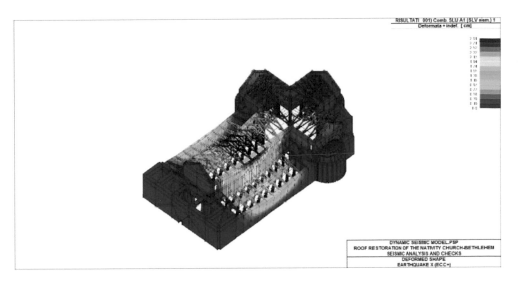

Figure 3.8.28 Deformed configuration for actions in the X (N–S) direction (plywood system).

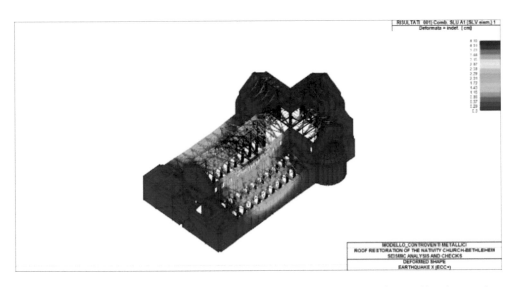

Figure 3.8.29 Deformed configuration for actions in the X (N–S) directions (crossed band system).

for the crossed band system. In particular, with the plywood system the maximum displacements at the higher and lower level are 2.91 cm and 1.95 cm respectively, whereas the same displacements with the largest steel bands available on the market (40 × 3 mm) become, respectively, 4.3 cm and 2.3 cm with an increment of 48% in the upper level and 21% in the lower level.

A further analysis was carried out by increasing the thickness of the plywood from 21 cm to 40 cm in order to consider as collaborating even the not-decayed part of the existing

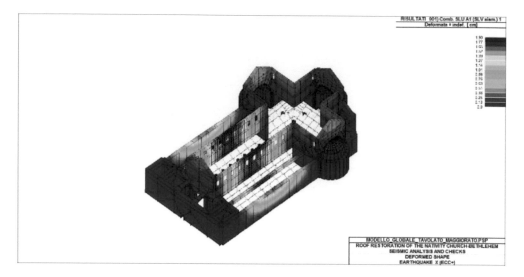

Figure 3.8.30 Deformed configuration for actions in the X (N–S) direction (increased plywood).

plank, i.e. 80% of its thickness. The stiffness of the plank is obviously increased as well as the boxlike behavior of the whole structure. Accordingly, the horizontal displacements in the X direction become lower and the seismic actions are better distributed among the walls (in Figure 3.8.30).

A comparison is based on the deformed patterns obtained: In Figure 3.8.30 the maximum displacement in the N–S direction (X direction) for the higher roof is 1.90 cm and for the lower roof it is 0.94 cm for 40 mm thickness, while for 21 mm thickness these values were, respectively, 2.91 and 1.95 cm with a reduction of 65% for the higher roof and 48% for the roof over lateral naves. All the verifications carried out confirm an improvement of the response of the structure to the expected seismic actions, defined in a return period of nearly two thousand years.

References

[1]   NTC – *Norme Tecniche per le costruzioni* – D.M. 14 Gennaio 2018.
[2]   Eurocode 8: *Seismic Design of Buildings*.
[3]   PRO_SAP *Professional Structural Analysis Program*, 2SI, Software e Servizi per l'Ingegneria, Ferrara, Italy. Build 2013.11.0167-RY2014(a).

3.9   Old and new roof covering

C. Alessandri and M. Martinelli

More difficult was to identify the date of previous interventions on the lead sheets of the covering structure. There are no sources that say unequivocally how the roof of the church was at the time of Justinian or in the centuries immediately following, even if the findings of excavations made at the beginning of the last century, as well as those relating to the most recent excavations, seem to confirm the presence of a tiled roof, as was customary in Byzantine constructions. As reported in [1][2], the first testimony is the one of the 11th-century pilgrim Hyacinth, who, in his description of the inside, also refers to a roof covered on the exterior with lead. A historically dated intervention was certainly that performed in the second half of the 15th century, when the Franciscan Guardian Giovanni Tomacelli obtained the sultan's permission for a thorough restoration of the roof, as mentioned by Fr Francesco Suriano and also witnessed by the original *firman* and Felix Fabri's account (see always [1] and [2]). In that occasion the king of England, Edward IV, provided the money for the lead covering. Since then the interventions of replacement were carried out over time with even greater frequency and, most of the times, without leaving any historic record. But unlike the wooden structures, at the time of the more recent survey campaigns, most of the lead sheets were seriously damaged and far below any acceptable standard of reliability, so as to make advisable their total replacement. Before their removal, however, all sheets were marked, and a detailed plan of their location within the entire covering was drawn (Figure 3.9.1). Moreover, after the removal, the sheets and most of the nails used for connections were stored in appropriate spaces.

It must be underlined that during the interventions carried out between the 18th and 19th centuries, the lead sheets had been covered by layers of bitumen and canvas impregnated of

Figure 3.9.1 Marked lead sheet.

Figure 3.9.2 Stored lead sheets.

bitumen as remedy to the many cracks that had been opened in the meantime, but in many parts of the roof, the bitumen had split because of the daily thermal excursions or melted under the high summer temperatures. The numerous traces of bitumen poured through the rainwater drainage system were evident (Figure 3.9.3) at the beginning of the restoration works.

The average thickness of the roof covering, including the bitumen layers, was 3.5 mm, but in some parts it was nearly 2 cm because of the deformations that occurred under the high temperatures of the hottest months and the material accumulation for gravity in the lowest parts of the roof (Figure 3.9.4).

It was clear that the existing roof covering was no longer suitable to guarantee the water-proofing of the roof and had to be removed completely. The removal of the lead sheets allowed the discovery of a layer of earth and straw inserted between the sheets and the wooden plank as thermal insulator (Figures 3.9.5, 3.9.6).

The composition of the whole existing roof covering, from the external bitumen layers to the internal purlins, is shown in Figure 3.9.7, while its weight per square meter is measured in Table 3.9.1.

Figure 3.9.3 Bitumen poured from drainage holes.

Figure 3.9.4 Thickness of the roof covering.

Figure 3.9.5 Earth and straw under the lead sheet.

Figure 3.9.6 Bitumen layers, lead sheet, earth/straw and planks.

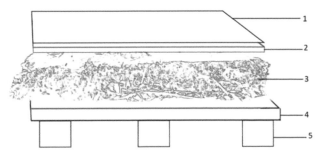

Figure 3.9.7 Roof covering composition.

Table 3.9.1 Roof covering weight per square meter

Previous roof covering	
Description	Weight/m²
1 Bitumen waterproof sheet	0.10 kN/m²
2 Lead sheets (3.5 mm)	0.40 kN/m²
3 Clay and straw (3 cm)	0.18 kN/m²
4 Existing boards (3 cm)	0.12 kN/m²
5 Purlins	0.27 kN/m²
Total	**1.07 kN/m²**

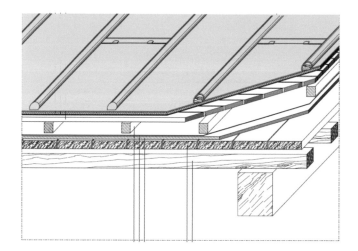

Figure 3.9.8 Layers of the new roof covering.

The new roof was conceived as a ventilated roof, made of different layers, in order to thermally insulate the innermost layers from the outer layers exposed to the highest temperatures and allow effective internal ventilation. In particular, starting from the bottom, there is a phenolic plywood layer F 15 (2 cm thick) just above the existing boards and strongly connected to them by screws (Figures 3.9.8, 3.9.9), a first vapor-control waterproof membrane, joists or wooden strips (40×60 mm) along the slope of the pitch to create the ventilation chamber, another layer of boards (2 cm thick and interspaced by a 5 cm gap; Figures 3.9.8, 3.9.10), a second waterproof membrane, a layer of sheep wool (high density: 100 kg/m³, 1 cm thick) to better isolate the underlying layers and prevent condensations and finally the lead sheets (2 mm thick; Figures 3.9.8, 3.9.11, 3.9.12, 3.9.13).

Figures 3.9.14 shows the shop-drawings of ventilation chamber and ridge, while Figures 3.9.15 and 3.9.16 show assembling phases of the ridge.

In the end (Figure 3.9.17) the life-line along the whole ridge of nave, transept and apse, which allows maintenance works on the roof by going up from the terraces of transept and apse, as shown in Figure 3.9.18.

The new roof covering has a weight per square meter (Table 3.9.2) less than that of the old roof (Table 3.9.1) but nevertheless meets all prescribed requirements of functionality and

Figure 3.9.9 Plywood layer.

Figure 3.9.10 First waterproof membrane, joists and boards.

Figure 3.9.11 Sheep wool layer.

Figure 3.9.12 Final lead sheets

Figure 3.9.13 Final lead sheets.

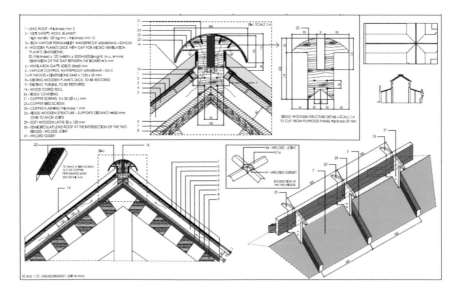

Figure 3.9.14 Ventilation chamber and ridge.

Figure 3.9.15 Assembling phases of the ridge.

Figure 3.9.16 Assembling phases of the ridge.

Figure 3.9.17 Life-line.

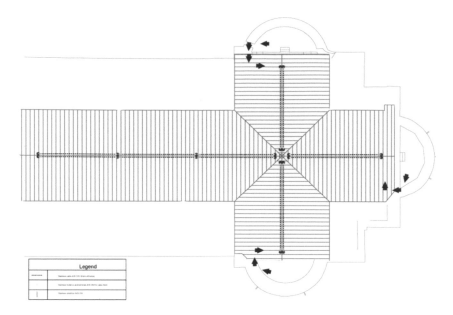

Figure 3.9.18 Life-line and access to the roof from the terraces.

Table 3.9.2 New roof covering weight per square meter

Secondary frame (purlins)	0.264 kN/m²
Existing boards (2,5 cm)	0.113 kN/m²
Phenolic plywood (2 cm)	0.104 kN/m²
Wooden joists	0.025 kN/m²
Boards supporting the wool layers (2 cm)	0.086 kN/m²
Wool layer (1 cm)	0.010 kN/m²
Lead sheets (2 mm)	0.230 kN/m²
Wooden rolls (for sheet connection)	0.017 kN/m²
Total	**0.849 kN/m²**

at the same time allows preserving the external appearance of the original roof. Moreover, this new covering, strongly connected to the existing wooden structures below, contributes to increase considerably the stiffness of the whole construction against seismic actions. All technical checks which were carried out to verify the water resistance of the new roof covering were successful. Moreover, the rainfalls which occurred since the completion of the roof left no traces inside. The restoration of the roof (1625 m²) was completed in March 2015 with the positioning of the new windows, a new life-line and new crosses.

It is worth noting that the design of the roof covering with lead sheets was drafted, where possible, with reference to the British Standard BS EN12588: 2006 [3] and to "Rolled Lead Sheet – The Complete Manual" [4], which is actually the corresponding implementation manual. Where, for various reasons (mainly linked to the importance and uniqueness of the building) it was not possible to follow these standards slavishly, some changes were made that in no way affect the effectiveness of the intervention.

The most relevant discrepancy is in the choice of 2 mm thickness for sheets, not compliant with the standards. This non-compliance is however due to a design choice of structural type made to reduce the seismic masses in the roof covering. The use of a non-compliant thickness implied a change in the size of the lead sheets (and therefore in the spacing of joists and clips) to ensure them adequate thermal expansion.

In fact, one of the main problems that can be encountered in designing metal roof coverings in general and lead coverings in particular concerns the correct sizing and positioning of the expansion joints. This problem is even more delicate and important in areas, such as those in the Middle East, where it is possible to have high thermal excursions over 24 hours.

For this reason, the geometry and the distribution of the expansion joints were carefully studied, both for the roof covering and for the gutters placed along the lower edges of aisles and corners. In particular, the gutters had to respect some existing constraints linked to the slope of the pitches and the location of the original drainage holes placed along the perimeter walls, both on the Orthodox side and on the Franciscan side, with a spacing from 2.7 m to 3.3 m between them (Figure 3.9.19).

Because of such constraints it was not possible to strictly comply with the BS EN12588 Standards. However, every effort was made to get solutions as close as possible to the law requirements and to allow maximum efficiency. In particular, the gutter joints were placed at a distance from each other corresponding to the maximum length that a lead sheet can have in order to avoid problems due to thermal expansion. The presence of such joints along the gutter is highlighted by ribs (Figure 3.9.20), the shape of which is similar to the one used for the joints of the lead sheets in the rest of the roof covering. Their distribution along the perimeter of the roof follows approximately the distribution of the cracks that occurred in the previous gutter in consequence of the high daily thermal excursions. As a matter of fact, these cracks occurred at a variable distance between 2,80 m and 5,00 m to release the high

Figure 3.9.19 Original drainage holes.

Figure 3.9.20 Ribs as expansion joints in the gutter.

internal stresses produced by hindered thermal expansions. Therefore, they can be seen as a natural adaptation of the roof covering to the local thermal excursions, as if the structure itself had created the expansion joints that had not been foreseen during the construction. It is reasonable to think that the new expansion joints distant from each other around 3.00 m (the minimum length allowed by the existing and non-modifiable geometry of the roof) can ensure a thermal expansion enough to prevent the occurrence of cracks in the gutter.

References

[1]   Bacci, M., Bianchi, G., Campana, S. & Fichera, G. (2012) Historical and archaeological analysis of the Church of the Nativity. *Journal of Cultural Heritage*, 13, Elsevier, e5–e26.
[2]   Bacci, M. (2017) *The Mystic Cave: A History of the Nativity Church in Bethlehem*. Rome, Brno. Viella/Masaryk University Press.
[3]   BS EN 12588:2006, Lead and lead alloys: Rolled lead sheet for building purposes.
[4]   Rolled lead sheet: The complete manual. Available from: www.leadsheet.co.uk

3.10 Windows

S. Palanti

The windows, dating back to the 19th century, are placed on three different levels: in the middle of each tympanum of apse and transept, at an intermediate level along the walls of nave and transept and at ground level in north and south transepts and apse. The inspection of their state of conservation started by the end of 2010 during the first survey campaign [1]. The analyses carried out on some samples taken from the wooden frames of the nave showed that they are made of European wooden species, such as *Pinus sylvestris* (Scots pine) and *Picea abies* (Norway spruce), both non-durable species in terms of biotic decay [2].

They were painted grey, but the paint was almost everywhere detached from the support because of the wood shrinkage and swelling caused by daily and seasonal hygrometric variations (Figure 3.10.1).

This weathering effect was more pronounced in the south side of the church because of the longer exposure of this side to the sun that had caused a visible warping (Figure 3.10.2).

In many cases the wooden frame had lost its perfect anchorage to the wall because of lack or loosening of mortar, and rainwater could enter inside very easily and cause serious damages to plasters and mosaics (Figure 3.10.3).

Also some panes were broken (Figure 3.10.4), and fungal decay had occurred in many parts (Figure 3.10.5).

Therefore, given the poor state of conservation of almost all the windows on the upper level, it was decided to replace them with new windows. On the contrary, the windows at ground level (Figure 3.10.6), more sheltered from wind, sun and rain, were undoubtedly in a much better state of preservation and needed only ordinary maintenance.

Figure 3.10.1 Decay of the painted surface.

Figure 3.10.2 Warping of wooden elements.

Figure 3.10.3 Loss of anchorage to the wall.

Figure 3.10.4 Broken glass.

Figure 3.10.5 Examples of fungal decay.

Figure 3.10.6 Window at ground level.

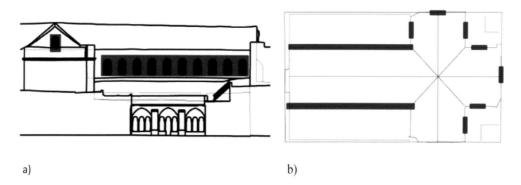

a)   b)

Figures 3.10.7 Windows on the upper level – (a) lateral view, (b) plan view.

All the 42 windows on the upper level (Figures 3.10.7a, b) were replaced with new windows having cypress frames and low-emissivity double glazing with special specifications to reduce the lighting in the church, on the wall mosaics and other decorated surfaces. The new windows were produced in Italy and brought to the site (Figures 3.10.8, 3.10.9). Of the original windows, they retain the same geometry and the same partition into mobile and fixed parts. The new dark-brown colour was chosen in agreement with the three religious Communities.

Figure 3.10.8 Details of the new window.

Figure 3.10.9 New windows on the south wall.

References

[1]   Ferrara Consortium (2011) *Final Report*.
[2]   European Committee for Standardization EN 350-2 (1997) Durability of Wood and Wood-based products. Natural Durability of Solid Wood, Guide to Natural Durability and Treatability of Selected Wood Species of Importance in Europe.

3.11   Maintenance

C. Alessandri, N. Macchioni, M. Martinelli and B. Pizzo

Premise

The recommendations reported in the present section are part of the maintenance plan of structures, which should accompany the various interventions on the church as a complementary document to the structural project, containing plans, programs and all the maintenance operations necessary to maintain functionality, quality features, efficiency over time and the economic value of the building after its complete restoration.

As prescribed by the Italian standards, Section 10.1 [1], the maintenance plan of structures consists of the following operational documents:

- the user manual;
- the maintenance manual

and it refers only to the structural elements of the building. However, the document is a dynamic instrument because it is the responsibility of the manager/owner of the building to keep it updated with all the documentation relating to the maintenance operations that are actually carried out.

User manual

The user manual refers to the use of the most important parts of the building, with particular reference to those parts that may generate risks due to incorrect use. It includes information on the location of the parts that have undergone interventions, their graphic representation, descriptions and methods of correct use.

Maintenance manual

The maintenance manual refers to the maintenance of the most important parts of the building. It contains the minimum acceptable level of performance, the anomalies that can be found, the maintenance operations that can be performed directly by the user and those that must be carried out by specialized personnel.

It also establishes maintenance and checks to be carried out following deadlines set in advance or following exceptional events.

The maintenance plan here presented refers only to the various components of the roof and to their connection with the walls.

3.11.1   Lead sheets

User manual

Prescriptions refer to the external elements of the roof covering directly exposed to weathering. They are new lead sheets (2 mm thick) that have been mostly worked on site (Figures 3.11.1, 3.11.2).

The correct use of these elements provides the preservation of the protection function from atmospheric agents of the structures and of the decorated surfaces of the church.

Figure 3.11.1 Lead sheets roof covering.

Figure 3.11.2 Lead sheet worked on site.

Maintenance manual

> *Minimum level of performance*: resistance to design stresses and atmospheric agents.
>
> *Abnormalities that can be found*: lead oxidation due to exposure to atmospheric agents, ripples and deformations due to failure and/or excessive deformation of the surface, deformations and breakages due to impeded thermal expansion, lifting due to atmospheric actions (wind, temperature), fractures at the welds, fillings and blockages due to dirt or materials carried by the wind.
>
> *Controls*: a first check is performed through a visual inspection aimed at identifying particular signs of deterioration. Later, if the element appears degraded, a more in-depth analysis should be carried out to identify the causes and define the repair interventions.
>
> *Periodicity of checks*: visual inspection every year and other specific analyses to be carried out when necessary. Visual inspection always carried out in case of particularly violent atmospheric events such as rains, snowfalls, windstorms.
>
> *Carrying out checks*: visual inspection is performed by specially trained personnel, because the material is very delicate and must be handled and treated following precise protocols; possible analyses must also be performed by specialized personnel.
>
> *Interventions*: they can vary depending on the problem and can range from simple cleaning and needed protective treatment to the replacement of the sheets or the expansion joints and/or welds.
>
> *Periodicity of the interventions*: they are not planned in advance but are carried out whenever necessary.
>
> *Performing interventions*: they must be done by personnel of appropriate competence.

3.11.2   Eaves and rainwater drainage system

User manual

These are the elements suitable for the collection and removal of rainwater (Figures 3.11.3–3.11.5). Especially in the case of low roof (aisles) it is essential that they are checked and maintained in perfect working order because their malfunctioning or failure would cause very serious damage to the supporting structures of the roof (walls and timber), plasters and wall mosaics.

The correct use of these elements requires the maintenance of the rainwater collection and removal function according to the project specifications, and these features should not be altered over time.

Maintenance manual

> *Minimum level of performance*: resistance to design stresses and atmospheric agents.
>
> *Abnormalities that can be found*: damage and oxidation due to exposure to atmospheric agents, ripples and deformations due to failure and/or excessive deformation of the surface, deformations and breakages due to impeded thermal expansion, lifting due to atmospheric actions (wind, temperature), fractures at the welds, fillings and blockages due to dirt or materials carried by the wind.
>
> *Controls*: a first check is performed through a visual inspection aimed at identifying particular signs of deterioration. Later, if the element appears degraded, a more

Figure 3.11.3 Drainage system in the corners.

Figure 3.11.4 Drainage system in the north aisle.

in-depth analysis should be carried out to identify the causes and define the repair interventions.

Periodicity of checks: visual inspection every six months and other specific analyses to be carried out when necessary. Visual inspection always carried out in case of particularly violent atmospheric events such as rains, snowfalls, windstorms.

Carrying out checks: visual inspection is performed by specially trained personnel, because the material is very delicate and must be handled and treated following precise protocols; possible analyses must be performed by specialized personnel.

Figure 3.11.5 Drainage system in the south aisle.

Interventions: they can vary depending on the problem and can range from simple cleaning and needed protective treatment to the replacement of eave parts or of expansion
joints and/or welds.

Periodicity of the interventions: whenever necessary.

Performing interventions: they must be done by personnel of appropriate competence.

3.11.3   Flashings and ridge elements

User manual

They are all the elements suitable for the protection of the terminal parts of the roof and
the connection of the roof covering to the church walls (Figures 3.11.6–3.11.9). Perfect
maintenance of these parts is essential because their malfunctioning or breakage can cause
very serious damage to the structures of the roof, plasters and wall decorations. In addition
to preventing water leakage in correspondence to the masonry walls, these elements have
the fundamental task of ensuring the ventilation inside the roof covering and preventing the
occurrence of condensation on its intrados. The condensation, in the long run, could cause
oxidation of the lead and especially rot, both in the wooden boards supporting the lead sheets
and in timber structures of the roof.

In order to be effective such elements must guarantee both the connection and the protection of the end parts of the roof covering and the internal ventilation (as specified in the
project), and these features should not be altered over time.

Figure 3.11.6 Flashing between aisle pitch and church counter-façade.

Figure 3.11.7 Flashing between aisle pitch and upper wall of the nave.

Figure 3.11.8 Flashing along the eave.

Figure 3.11.9 Ridge of the roof.

Maintenance manual

Minimum level of performance: resistance to design stresses and atmospheric agents.

Abnormalities that can be found: damage or oxidation due to exposure to atmospheric agents, ripples and deformations due to failure and/or excessive deformation of the surface, deformations and breakages due to impeded thermal expansion, lifting due to atmospheric actions (wind, temperature), fractures at the welds, fillings and blockages due to dirt or materials carried by the wind.

Controls: a first check is performed through a visual inspection aimed at identifying particular signs of deterioration. Later, if the element appears degraded, a more in-depth analysis should be carried out to identify the causes and define the repair interventions.

Periodicity of checks: visual inspection every year and other specific analyses to be carried out when necessary. Visual inspection always carried out in case of particularly violent atmospheric events such as rains, snowfalls, windstorms.

Carrying out checks: visual inspection is performed by specially trained personnel, because the material is very delicate and must be handled and treated following precise protocols; possible analyses must be performed by specialized personnel.

Interventions: they can vary depending on the problem and can range from simple cleaning and needed protective treatment to the replacement of some parts or of expansion joints and/or welds.

Periodicity of the interventions: whenever necessary.

Performing interventions: they must be done by personnel of appropriate competence.

3.11.4 Substructure (boards, joists, plywood)

User manual

These elements support the lead sheets of the roof covering. Starting from the extrados the whole package (Figure 3.11.10) is composed of sheep's wool mat (10 mm thick) immediately

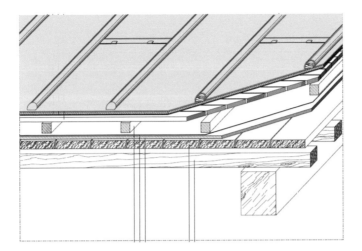

Figure 3.11.10 Cross section of the ventilated roof covering.

under the lead sheets, waterproof membrane, fir boards (20 mm thick), small spruce joists for the ventilation chamber (transverse section 40 × 60 mm), waterproof membrane, phenolic plywood (20 mm thick). Maintaining these parts of the roof in perfect working order is essential because their malfunctioning or breakage can cause very serious damage both to the lead coat and to the bearing structures of the roof and consequently to the whole church. In addition to bearing the weight of the lead sheets, these elements have the fundamental task of ensuring ventilation, preventing the occurrence of condensation on the inside of the covering package. In the long run, condensation could cause lead oxidation and especially rot due to fungal attack, causing the decay of the internal bearing structure of the roof covering.

The correct use of these elements requires the maintenance of the support function of the wooden boards and the ventilation of the lead sheets according to the design specifications, and these features should not be altered over time.

Maintenance manual

Minimum level of performance: resistance to design stresses.

Detectable anomalies: attack by insects or by fungi in case of percolation of water or permanence of high levels of humidity for a long time.

Controls: as the various elements are invisible from the outside, the possible anomalies cannot be directly detected. A first check is carried out through visual inspection aimed at identifying particular signs of degradation (such as traces of water leaks in the lead). During the visual examination, particular attention must be paid to the identification of phenomena such as local deformations, excessive deformations, incipient breaking, differential displacements between the various elements etc., which may indicate a possible problem within the coverage package. Later, if the element appears degraded, a more in-depth analysis must be carried out, including instrumental (endoscopic, thermographic etc.) analyses, to identify the causes and define repair interventions.

Frequency of checks: visual inspection every year and other specific analyses to be carried out when necessary. Visual inspection always carried out in case of particularly
violent atmospheric events such as rain, snowfall, windstorms.

Carrying out checks: the visual check of the lead covering is carried out by specifically
trained personnel, because the material is very delicate and must be handled and
treated following precise protocols; the other possible analyses involving the underneath layers must be carried out by specialized personnel.

Interventions: they must be carried out by specialized personnel, depending on the problems encountered. They include protective treatments of elements slightly attacked
by insects (in order to prevent the infestation from spreading to healthy material and
hence the damage progression in the attacked wood) to the replacement of particularly decayed parts. Substitutions must always be made based on a project drawn up
by qualified and experienced technicians.

Frequency of interventions: whenever it is deemed necessary.

Carrying out of the interventions: they must be done by personnel with suitable competence.

3.11.5   Wooden structure (trusses, cantilevers, purlins, timber plank)

User manual

It refers to the main and secondary timber elements (Figure 3.11.11) that make up the bearing structure of the roof, their connections and the elements connecting the wooden structures to the masonry walls.

Figure 3.11.11 Some of the timber elements of the roof structures.

The timber elements are in old solid wood of various species, mainly oak, cedar and larch. The original connections were made with nails and brackets in forged iron, whereas in those repaired in the last intervention campaign some modern materials have also been used, such as galvanized or stainless steel screws, galvanized or stainless threaded bars and epoxy resins. When it was necessary to disassemble the existing connections to carry out interventions, the original nails were preferably reused after being straightened and checked in quality and integrity.

The correct use of the elements requires the maintenance of the operating loads within the project limits and their non-alteration over time, the preservation of the effectiveness of the joints between original elements and prostheses, a wood moisture compatible with the use class 2 according to EN 335:2013 [2], corresponding to situations in which wood is not exposed to the weather, but occasional, although not persistent, wetting can occur.

Maintenance manual

Minimum level of performance: resistance to design stresses.

Detectable anomalies: weathering due to exposure to atmospheric agents (for the external elements), attack by insects, rots due to percolation of water and permanence of high levels of humidity for a long time. The most important causes of decay recorded during the diagnosis campaign were fungal rots and subterranean termite attacks. Both occur in consequence of a high wood moisture content: fungal rot needs wood moisture content higher than 20 to 25% during, at least, some weeks, and subterranean termites usually attack wood that remains wet for prolonged periods and cause damage that is generally severe and difficult to identify. Those two categories of decay agents in wooden elements are the most important in all timber structures, and their importance was confirmed in an only apparently drier climate such as the one of the church of the Nativity.

Insects that degrade dry wood (principally by *Anobidae* and *Cerambicidae* families) can be considered an actual danger only for the new products installed during the restoration, more particularly for the wooden planks placed between the lead cover and the supporting structure. In fact, the timber elements installed during the restoration works were also old, and they had a limited amount of sapwood. Very hardly possible attacks will be transferred to the main wooden structure, where, during the diagnostic investigation, all the attacks by this type of insect were no longer active.

In order to reduce the risk of attacks by fungal rot, the utmost care must be taken in assessing the presence of possible moisture traps. During the previous diagnostic investigation, these had clearly developed in the interface between the masonry and the wood due to the water that penetrated from the roof and accumulated inside the walls.

Controls: a first check is carried out through a visual inspection aimed at identifying particular signs of degradation. The checks are carried out by measuring the wood moisture contents as close as possible to the wooden portions in contact with the masonry, in particular 1 to 2 weeks after rainy events. As an alternative to direct investigations, a more in-depth analysis can be carried out by installing some electronic devices that allow remote measurements. If the wood moisture measurements carried out at 4 to 5 cm depth prove to be potentially dangerous for the wood, the causes of the event must be found as quickly as possible in order to remedy. In fact, even if the wood dries, without remedying the cause of the accumulation of moisture, this would show up again.

It is therefore very important that the use conditions of class 2 (as defined in EN 335; see above) are actually maintained for all the portions of the wooden structure. Those most at risk of moving towards conditions 3.1 (wood does not remain wet for long periods. Water does not accumulate) and 3.2 (wood does remain wet for long periods. Water may accumulate) are certainly those in contact with the masonry.

Frequency of checks: visual inspection every year and other specific analyses to be carried out when necessary. In particular, the assessment of the presence of termite attacks can be carried out at least 1 to 2 times a year. The underground termites in fact make their nest in the ground and reach the woody material they feed on, bringing it inside the nest itself, through artificial tunnels easily recognizable by expert personnel. Also in this case the maintenance of wood in a sufficiently dry state can be considered a form of prevention, although not complete.

Carrying out checks: the visual inspection can be performed by the trained user, whereas the other analyses must be carried out by specialized personnel.

Interventions: they must be carried out by specialized personnel, depending on the problems encountered. They include protective treatments of elements slightly attacked by insects, in order to prevent the infestation from spreading to healthy material and hence the damage progression in the attacked wood. Use of preservatives as impregnating (and not filmable) products.

Frequency of interventions: whenever it is necessary. Preservatives against insects should be applied as a minimum, after 3 years the first time, after 5 years the second time and every 10 years the subsequent times.

Carrying out of the interventions: must be performed by specialized personnel.

3.11.6   Connectors

User manual

They are all those elements in steel, like nails, bolts, pins and screws, suitable for joining two or more wooden elements and transferring mutual forces (Figures 3.11.12–3.11.14). The durability of such elements is guaranteed by a galvanizing process, which limits the occurrence of micro-electrolytic cells with anodic action.

The correct use of these elements requires the maintenance of the operating loads within the project limits and their non-alteration over time.

Maintenance manual

Minimum level of performance: resistance to design stresses.

Detectable anomalies: bubbles or cracking of the protective layer with risk of corrosion (chemical and biological causes). Partial or total loosening of the connector with possible exit from their own seat. Connector deformation, deformation of the connected elements due to connector's transversal expansion.

Controls: they are carried out through a visual inspection aimed at identifying the possibly damaged item.

Frequency of checks: every year.

Carrying out checks: they can be performed by the trained user.

Interventions: application of anti-rust products thus to restore the protective layer, or replacement of the damaged element.

Figure 3.11.12 Some new boards of the roof covering where screws were used.

Figure 3.11.13 Steel connector between truss end and wall.

Figure 3.11.14 Steel connector between purlins/boards and wall.

Frequency of interventions: whenever it is necessary.

Carrying out of the interventions: concerning the restoration of the protective layer, they can be performed by the trained user. The substitution of damaged elements must be done by personnel with suitable competence.

3.11.7   Metal carpentry

User manual

They are all those elements in steel or forged iron suitable for joining two or more elements and transferring mutual forces. The durability of newly installed elements is guaranteed by a galvanizing process, which limits the occurrence of micro-electrolytic cells with anodic action.

The correct use of these elements requires the maintenance of operating loads within the project limits and their non-alteration over time.

Maintenance manual

> *Minimum level of performance*: resistance to design stresses.
>
> *Detectable anomalies*: bubbles or cracking of the protective layer with risk of corrosion (chemical and biological causes). Partial or total loosening of the connector with possible exit from their own seat. Element deformation, deformation of the connected elements due to connector's transversal expansion.
>
> *Controls:* they are carried out through a visual inspection aimed at identifying the possibly damaged item.
>
> *Frequency of checks*: every year.
>
> *Carrying out checks*: they can be performed by the trained user.
>
> *Interventions:* application of anti-rust products to restore the protective layer or replacement of the damaged element.
>
> *Frequency of interventions*: whenever it is necessary.
>
> *Carrying out of the interventions*: concerning the restoration of the protective layer, they can be performed by the trained user. The substitution of damaged elements must be done by personnel with suitable competence.

3.11.8   Prostheses

User manual

They are all those wooden elements used to replace the decayed parts of the timber structure of the roof and are suitable for transferring mutual forces between the two parts (the original and the replaced one) of a same structural element. The prostheses, normally in old wood, are connected with chemical systems (epoxy resins used to glue some connecting metal bars) (Figures 3.11.15, 3.11.16) or mechanical ones (metal connectors with a

Figure 3.11.15 Glued-in rod.

Figure 3.11.16 Covering wooden strip.

cylindrical shank, such as bolts or pins or similar). In the case of prostheses with glued-in rods, the durability is guaranteed by the protection offered to steel bars by the resin and the wooden lath covering both bar and resin; of course, to preserve glued connections from delamination it is needed that moisture variations are limited in timber elements, this meaning that water leakages from the roof must be avoided (as specified above). In the case of prostheses with mechanical connection, the durability is ensured by the use of stainless steel connectors.

Moreover, the correct use of these elements requires the maintenance of operating loads within the project limits and their non-alteration over time.

Maintenance manual

Minimum level of performance: resistance to design stresses.

Detectable anomalies: delamination in the glued bars area, abnormal cracks near the glued bars, ejection and/or deformation of the metal pins in half-lap joints, and in general disconnection of metal connectors, deformation of the wooden elements due to wood embedment, detachment or rotation of the prosthesis.

Controls: they are carried out through a visual inspection aimed at identifying the possibly damaged item.

Frequency of checks: every year.

Carrying out checks: they can be performed by the trained user.

Interventions: if only one of the anomalies listed above is found, an immediate check must be carried out by a specialized professional to evaluate, even with instrumental investigations, the actual severity and danger of the anomalies identified. If necessary, safety measures must be taken, and static checks of the involved prosthesis must be carried out. Possible consolidation works should also be considered.

Frequency of interventions: whenever it is necessary.

Carrying out of the interventions: every needed structural intervention must be carried out by specialized personnel.

Remarks

In addition to what was already said before and to underline the importance of immediate checks in the case of some exceptional events the following recommendations are also provided.

After or during very heavy rains, it is necessary to carry out a global check of all the vertical and horizontal elements of the wood structure in order to verify possible infiltrations of rainwater. If necessary, a prompt intervention must be carried out for a rapid drying of elements, and in extreme cases suitable preservatives against the occurrence of fungi should be used.

In the case of exceptional windy events, all the connections at the base of the roof structural elements must be carefully checked, and their replacement must be immediately performed in case of serious damage. The perfect state of conservation of the roof covering and of its connection to the bearing structures must also be adequately controlled, and a prompt intervention must be performed in the presence of some damage that might cause water infiltration inside the covering package or on the wooden structures below. In this case, it is necessary to assess the possible state of decay of the wooden elements by means of inspections to be carried out by wood experts. If necessary, a prompt intervention must be planned, based on cleaning and removal of the decayed layers, consolidation or replacement of wood (in the most serious cases) and application of suitable products to prevent the occurrence of fungi.

If a seismic event occurs with a peak ground acceleration higher than the one considered in the design earthquake, it will be necessary to carry out a punctual check of all the connections between wooden structures and walls, of the joints among wooden elements and of the integrity of the prostheses. Damaged connection elements (nails, screws, metal plates) must be replaced by new elements placed next to the existing ones and at suitable distance; the same number and type of connection must be used.

For seismic events with a Richter magnitude greater than 4.5, corresponding to a Mercalli scale higher than IV, the damage degree of the structures must be verified by means of sample checks of the most stressed elements, such as

- Connections of the trusses to the masonry walls
- Rafter–tie beam connection
- King post–rafter and king post–tie beam connection
- Strut–rafter and strut–king post connection
- Glued prostheses
- Mechanical prostheses.

The truss supports and the ends of tie beams and rafters must be checked annually. In the case of rainwater infiltration it will be necessary to find out and remove the cause of infiltration, let the wet parts dry, clean and remove the decayed layers, consolidate the damaged part or replace it in the most serious cases and, if needed, apply suitable protection products against fungi.

References

[1]   NTC 2008 – Norme tecniche per le costruzioni – D.M. 14 Gennaio 2008.
[2]   EN 335:2013 *Durability of wood and wood-based products: Use classes: Definitions, application to solid wood and wood-based products.*

Chapter 4

Mosaics

4.1 Historical analysis

M. Bacci

The mosaic medium was chosen to embellish the walls of the Nativity Church in Bethlehem already during its reconstruction in the late sixth century, but nothing is known of its program. Apparently, some remnants of this original decoration were present in the early 12th century, as is witnessed by the Russian pilgrim Daniil's hint at them in his travelogue, and also by the fact that a number of older tesserae were integrated into the new mosaics. The latter were made under very specific political circumstances in the 1160s: as we are informed by the bilingual Greek and Latin inscription in the bema, the works were achieved in 1169 and were promoted and financed by the Byzantine Emperor Manuel Comnenos, the King of Jerusalem Amaury I, and the Latin Bishop of Bethlehem Raoul. The church was thoroughly reveted with an uninterrupted cycle of mosaics decorating the upper layers of the walls in the nave, the transepts and the main apse.

Exceptionally for the Middle Ages, we are very well informed about the authors of the program. The name of a certain "Master Ephraim" is given a prominent place in the dedicatory inscription, where emphasis is laid on his being not only a painter but also a "mosaicist", that is an expert in this specific technique. It can be assumed that he had the role of team's head and was responsible for the making of the images put on display on the north transept, where the strongest stylistic connections with contemporary Byzantine Comnenian painting can be detected. Other artists, whose names could still be read in a Greek inscription in the south transept in the early 17th century, dealt with the decoration of that area, though probably under the supervision of Ephraim. Another artist named *Basilius* included his name close to an angel's feet in the mosaic between the seventh and the eighth window on the north wall of the nave: as it can be inferred from his double signature in Syriac and Latin, he was probably an Arab Christian, that is an artist of local, Palestinian origins.

The decoration campaign in Bethlehem was probably the most ambitious in the whole history of the arts in the Latin Kingdom of Jerusalem. It can be assumed that the new works were stimulated by some damage that occurred in 1160, when a powerful earthquake destroyed the nearby monastery of Mar Elias. The involvement of the Byzantine Emperor as sponsor took place in the frame of political circumstances favouring the rapprochement of the Crusader Kings with Constantinople. The program, probably inspired by Bishop Raoul and his canons, stood out for its uniqueness and sophistication. On the one hand, it aimed to stimulate visitors to meditate on the mystery of Incarnation celebrated in the holy site: upon entering, a procession of angels on the upper walls of the central nave oriented the pilgrim's look towards the apse, where Mary was represented between Abraham and David. Pilgrims then entered the holy cave from the south transept, dominated by the *Nativity* scene in the south apse and several scenes from Christ's life, then made their devotions before the site of Christ's birth (also decorated with a *Nativity*) and the Holy Manger, and finally exited from the northern flight of steps to the north transept, where the full story of the Lord's Passion and Resurrection was put on display. When looking back toward west, they could remark that the full genealogy of Christ's ancestors according to both Matthew and Luke guided the observer, on both sides of the central nave, up to the majestic image of the *Tree of Jesse* on the counter-façade.

The program in the nave was dominated by a sequence of themes standing out for their being neither narrative nor iconic in character: they consisted of architectural frames housing altars and long inscriptions, which were meant to evoke the seven ecumenical councils on the south wall of the central nave and seven of the provincial synods of the early church on the northern one. Inspired by the synopses of the conciliar resolutions used by the Greek-rite or Melkite Arab Christian communities of Syria and Palestine, such images laid emphasis on the theological authority of the church in terms that could be shared by both the Latin and the Greek clergy.

The mosaics were meant to have a strong impact on beholders and, therefore, they were embellished with very prominent ornamental motifs. Fantastic, jewelled plants, inspired from the décors of the Dome of the Rock and the Aqsa Mosque in Jerusalem, were used to separate the symbolic representations of the councils. Between the latter and the upper layer with the procession of angels, the mosaicists inserted a narrower ornamental band with elegant tendrils, inhabited by different sorts of animals and precious objects as in contemporary Western miniatures, whereas other vegetal motifs displaying mushrooms were used to embellish the upper border.[1]

4.2   Materials and techniques

S. Sarmati, N. Santopuoli and E. Concina

4.2.1   Wall mosaics description

Mosaics decorated all the walls in the Church of the Nativity. The decorative cycle included a *Tree of Jesse* on the counter-façade, whereas the provincial councils of Palestine, the seven Ecumenical Councils, a lower band showing Christ's ancestors and an upper band between the windows with full-length images of angels were put on display in the central nave. There was a selection of Gospel scenes in the transept, a depiction of the Annunciation in the triumphal arch, and an image of the Virgin Platytera between Abraham and David in the apse.

There were most likely mosaic decorations on the narthex walls as well.[2]

The evocative and didactic program that can be reconstructed from both the material investigation of the extant mosaic surfaces and the analysis of historical sources aimed to accompany pilgrims on their journey towards the Nativity Grotto and orientate their physical experience of the Bethlehem holy site.

It can be easily guessed that the mosaic program, in its original state, with its uninterrupted sequence of mosaics made out of glass and gold tesserae and mother-of-pearl inlays, gave shape to a corridor of light through which visitors were guided toward the hallowed cave.

Unfortunately, only a few fragments remain of this complex decorative composition.

A surface area of about 127 square meters of the original mosaic remains, of which 92 are on the north and south walls of the nave (Tables 4.3.1, 4.3.2, 4.3.3). There are about 30.86 square meters mosaic in the north and south transepts (Tables 4.3.4, 4.3.5). Only 2.48 square meters mosaic remains in the north and south walls of the bema. In the south wall there is a dedicatory inscription with the names of those who commissioned the work, the person who executed the work and its date of completion.

In the past, the mosaic was supposed have been created in distinct phases at different times,[3] but now, according to analyses made during restoration, it seems certain that the entire decoration is homogeneous in both its technique and materials employed.

It is presumable, however, that many teams worked on the nave and transept at the same time with different operating procedures. In fact, there are differences in working methods as regards the size of the tesserae, the application and composition of bedding mortar, the tessellation of tesserae, the execution of figurative details, etc.

Most of the remaining original decoration is to be seen on the north wall of the nave (Figure 4.2.1).

In the upper register was depicted a procession of angels, of which seven remain (six were already visible, and the seventh was discovered beneath plaster during restoration). Each full-length figure stands within a set of windows. The latter are taller than the windows and stand out against a gold background made of tesserae embedded at an almost 45° angle to reflect more light.

Each figure is long and slender with a long tunic reaching the ankles and a *himation* that drapes over the left shoulder and arm, making a fold above the wrist. The figures' hair falls in

Figure 4.2.1 North wall mosaics.

Figure 4.2.2 The fourth and fifth angels on the north wall.

large curls on the neck. Different shades of stone tesserae compose the hair's dark colour, while transparent black glass tesserae outline the curls. Each figure has a ribbon in its hair shaping a kind of diadem on the forehead. The ends of the ribbon flutter into a U shape inside the aureole. Different shades of stone tesserae compose the faces and complexions, while transparent black tesserae outline eyebrows and eyes. A round transparent black tessera forms the iris.

The depicted figures are walking towards the transept, as if wanting to accompany pilgrims towards the grotto. They are portrayed in two alternating poses, making the image more dynamic: the first has a static pose with both feet planted on a flowery meadow; the second has a more dynamic pose in which the figure's left leg is raised in the act of moving forward (Figure 4.2.2).

A horizontal line of transparent black tesserae marks the horizon line separating the gold background from the meadow, made with opaque glass tesserae in various shades of green and yellow on which stand the angels. On the meadow, leaves and long-stemmed ovoid flowers are scattered, some with an unusual mushroom-like shape, outlined in light- and dark-blue tesserae and filled in with opaque green and yellow tesserae (Figure 4.2.3).

The wings are regularly made according to the same shape: one wing has a higher, diagonal position behind the head and halo, whereas the other has a vertical position on the figure's back.

The differences between each pair of angels are marked in the alternation of the wings, because one obtains its highlights from thin rows of silver tesserae, while the other angel is marked with rows of gold tesserae. A decorative band with a continuous plant volute of green leaves and red-capped mushrooms decorates the top of the mosaic.

Figure 4.2.3 North wall. The meadow at the feet of the second angel.

This is probably where the wooden ceiling had been mounted, presumably covering the wooden trusses of the roof of the church[4] (Figure 4.2.4).

A band of transparent black tesserae creates the vertical limit of each panel. During restoration, a geometric band of red triangles on white background was discovered around the arch of a window (Figure 4.2.5).

Depicted immediately below the band with angels, a taller band separates the upper register from the middle one, where the provincial councils of Palestine are. The compositional scheme of this frame is of great originality and grandeur. Wide plant volutes of leaves of different colours space out with small trilobed leaves, heart-shaped buds and more red-capped mushrooms. Posed between each plant volute are well-defined zoomorphic shapes such as hares, dromedaries, heads of dogs, birds and antelopes, as well as vases and chalices, showing us a great deal about the irony and mastery of the mosaicists (Figure 4.2.6).

Of the seven provincial councils, only those of Antioch and Serdica (with candelabra on the side) remain. The entire lower band is missing. However, we have an example of what the bottom register would look like by comparing the fragments of mosaic remaining on the south wall of the nave.

The councils on the north wall are religious buildings that, while stylised, are very well structured with upper galleries, domes and turrets covered with roof tiles. Though the structures are a product of pure fantasy, the depiction is realistic. A thin architrave held up by four columns marks each scene. Each of the Corinthian-style capitals consists of an abacus with a central loop, two sets of lateral volutes, a frontal caulicle and a very tall astragal. The column shafts vary in shape and colour. Curtains hang from each pair of columns and are either folded and draped or hooked tight across the architrave, in such a way as to make the

Figure 4.2.4 North wall. The decorative band at the top of the mosaic.

Figure 4.2.5 North wall. The geometric band discovered under the plaster.

interior visually inaccessible. Above the altar are a central arch and a long inscription on a gold or white background that refers to the council.

On the south wall of the nave, only three isolated fragments of mosaic remain. Starting on the left, the first fragment represents line endings from the First Council both sides. The second fragment shows approximately half of the Council of Ephesus and almost of the whole of the Council of Chalcedon. The third and final fragment on the far right contains a small piece of the upper part of the Third Council of Constantinople.

Figure 4.2.6 North wall. Antelope.

The Ecumenical Councils are represented with simpler images than those of the provincial councils on the north wall. Three columns with shafts that vary in decoration and colouring and topped with capitals are similar to those on the opposite wall with very prominent and colourful abacus and echinus. The columns support two arches, beneath which is a long inscription explaining the council and two altars seen in perspective, one flanked by ostensory and the other by candelabra.

The largest fragment on the far left side of the wall shows what the bottom register would have looked like throughout the nave. Immediately above the architraves, there is a row of half-length figures representing Christ's ancestors, with their names written to the right in transparent black tesserae on a gold background, separated from the middle register by a horizontal line of black tesserae (Figure 4.2.7).

A band with a black and white geometric motif marks the top of the wooden architrave finishing the decoration (Figure 4.2.8).

Of the long procession of ancestors, only seven remain. The inscription with their names can identify them. Starting from the left, we have Jacob, Matthan, Eleazer, Eliud, Achim, Sadoc and Azor. Each one is shown in a head-and-shoulder image that either faces forward or is slightly twisted. All have a halo made according to a concentric pattern, which can be gold or filled with blue or green glass tesserae or of dark red stone tesserae. Almost all of them are wearing a tunic and himation. Their hands are in the act of blessing or holding scrolls. Transparent black glass tesserae outline the faces, hair and beards, as their eyes and eyebrows. Round, transparent black glass tesserae outline irises just like the angels on the north wall. A line of red glass opaque tesserae and a line of pink stone tesserae outline the nose and mouth.

Figure 4.2.7 South wall. Christ's ancestors.

Figure 4.2.8 South wall. Christ's ancestors.

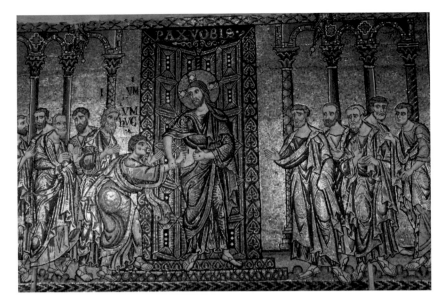

Figure 4.2.9 North transept. The incredulity of St. Thomas.

There are only a few remaining fragments of the original decoration in the transept and two strips in the bema.

On the east wall of the north transept, we can still see the representation of the *Incredulity of Thomas* as well as the bottom portion of the *Ascension* (Figure 4.2.9, 4.2.10).

The first scene is the best preserved. In the middle is Christ, with one group of apostles, including Saint Thomas, on his right and another group of apostles on his left. A simple architectural structure frames the figures: columns with Corinthian capitals support a series of arches. Behind Christ stands a slightly open door.

In both scenes, the full-length figures are standing on a meadow of light green, almost yellow glass tesserae with trilobed flowers outlined in dark-blue glass tesserae and filled in with various shades of light-blue glass tesserae and light-blue and white stone tesserae. They are all wearing light sandals, like those of the angels, made with a row of black tesserae that simulates light soles and straps wrapping the figure's big toe and knot on the side and forming either a round or triangular eyelet.

Only the lower portion of the *Ascension* remains. The scheme is classical; the Virgin is in the centre and on each side are an angel and a group of apostles. Figures make theatrical gestures creating an effect of movement. The apostles, without aureoles, are looking upwards and depicted in profile or three-quarter view (Figure 4.2.11).

The angels beside the Virgin have gold aureoles made of glass tesserae, arranged in concentric circles, unlike the aureoles of the angels on the north wall whose tesserae are partially in parallel lines. The angels in this scene also have ribbons in their hair that form a sort of diadem on their foreheads and whose ends flutter around the haloes, demonstrating the use of a figurative scheme that had been established and then carried out in a different form. The figures' complexions are made of stone tesserae in various shades of pink and outlined by a row of red glass tesserae. Transparent black glass tesserae outline their hair, eyes and eyebrows. A single, round, transparent black glass tessera shapes each iris.

Figure 4.2.10 North transept. The Ascension.

Figure 4.2.11 North transept. The Ascension, detail.

A lozenge frame made of gold tesserae on a bright red background delimits the two scenes laterally. The frame at the base of the scene is made of blue and red arrows on a gold background (Figure 4.2.12).

Below the two panels is a wide frame alternating triangle and acanthus leaf motifs on a gold background. The compositional scheme of the frame is of great originality and grandeur. Black glass tesserae highlight the perimeters of the leaves and geometrical figures. There are

Figure 4.2.12 North transept. The lozenge frame.

Figure 4.2.13 North transept. The frame below the mosaic panels.

mother-of-pearl circles and lozenges on top of the triangles. This wide band extends to the edge of the wooden architrave (Figure 4.2.13).

On the east wall of the south transept, we can still see a small portion of the *Transfiguration* with Saint Peter, the Apostle and the *Entry into Jerusalem* (Figure 4.2.14).

Figure 4.2.14 South transept. The Transfiguration.

Only the lower part of the *Transfiguration* remains, showing the figure of an apostle, probably Saint Peter,[5] falling down on a side as he is struck by a beam of light made of parallel rows of silver glass tesserae. In this case, the apostle has a gold halo with tesserae arranged in concentric circles and an outline marked with black glass tesserae. The face is rendered with transparent black glass tesserae, which give this figure a distinctive appearance if compared to the apostles' faces in the north transept, whereas the arms and legs are marked by the same row of red glass tesserae. The figure stands out on an arid hill, made of ochre stone tesserae and parallel lines of opaque green and yellow glass tesserae. There is a small bush on each side of the figure; the one on the left is contoured in black glass tesserae, small blue and green needle-shaped leaves and teardrop-shaped flowers made of gold tesserae. The one on the right has a row of red glass tesserae around the edges and round mother-of-pearl fruits. The frame around the panel has a geometric motif composed of rounds and lozenges on a red background, being close to the ones we find around the angels in the nave, where the background is white.

The second panel with the depiction of the *Entry into Jerusalem* is more complete, although only the top portion is missing (Figure 4.2.15).

This scene repeats a consolidated representation scheme. On the left is depicted Christ riding a donkey accompanied by Saint Peter. On the right, there is a series of people coming to meet them with the high walls, towers and houses of Jerusalem at their backs.

A bendy palm tree that reaches the top of the panel with two young boys climbing on it completes the scene. There are two more boys laying garments at the feet of the Redeemer. Transparent black glass tesserae outline all the faces, except for those of the children, depicted in profile or three-quarter view outlined with red glass tesserae.

Christ and Saint Peter have gold aureoles contoured in black tesserae. The gold tesserae are slightly angled downward so as to reflect more light, and they are laid in horizontal

Figure 4.2.15 South transept the Entry into Jerusalem.

Figure 4.2.16 South transept. The Christ.

rows and are slightly larger than the one used in the background, probably to make them look more important. The three arms of Christ's cruciform aureole are decorated with large mother-of-pearl rhombuses in the centre.

The figures of Christ and Saint Peter seem to have larger dimensions than the other figures to focus attention on them. Long rows of gold tesserae decorate Christ's *palio* in order to make him stand out of the scene (Figure 4.2.16).

The frame around this panel also has a geometric motif composed of rounds and lozenges on a red background, like the one seen on the north wall of the nave.

In this scene, we can see horizontally and vertically the construction lines of the end of the *pontata*.[6] The two panels are connected to the architrave with a wide band decorated with plant motifs in gold and mother-of-pearl tesserae on a green background (Figure 4.2.17).

The mosaics in the nave differ from those in the transept in tesserae size, bedding technique and other technical details. As previously mentioned, these differences are surely due to the various teams of mosaicists working together at the same time.

Understanding how a mosaic has been laid and finding out the sequence in which the work has been performed is essentially a deductive process. After careful observation, we can observe that the work probably started from the top. The figures and decorative motifs were first outlined and then filled in. The backgrounds made with gold tesserae were laid last. A row of gold tesserae runs around the perimeter of the figures and other decorative elements to indicate the areas to be filled in with gold tesserae (Figure 4.2.18).

Figure 4.2.17 The red arrows pointed out the joint of vertical pontata; the black arrow shows the horizontal one.

Figure 4.2.18 North wall, detail.

4.2.2   Execution technique

The masonry support for the mosaic is made of large square-cut stone. To prepare the support, the blocks were first lightly chiselled so that the plaster would adhere better. Nails, which were used to reinforce the anchoring of the plaster to the wall, were found in some of the indentations in the blocks (Figure 4.2.19).

They studied the sequence of preparatory layers both in the nave, where the edges of various mosaic fragments were exposed, and in the transept, where we found a large portion of supporting plaster applied to the wall in the north bema.

Figure 4.2.19 North wall, detail of the cuts on the stone under the mosaic.

It seems likely that the wall had been covered with a base layer of plaster made of lime and fine inert material on top of which the preparatory sketches were executed. In the support plaster seen in the north bema we can clearly see the *battitura* of wires, preparatory drawing techniques, executed in red ochre (Figure 4.2.20), as well as the horizontal incisions made with cardboard (the edges are rounded) as a guide for the application of the setting bed for a geometrical decoration.

Here, the base layer is about 2 centimetres thick, whereas in the nave it varies from just under 2 centimetres to much thicker than 2 centimetres, probably in order to level out the uneven surface of the wall.

The next layer, the setting bed, is composed of lime without traditional fillers. It includes, however, pieces of straw, which are useful in slowing down the drying process so that there is more time to apply the tesserae. This layer is 2 to 3 centimetres thick (Figure 4.2.21).

The setting bed was regularly fresco painted, in such a way as to be used as a guide for the application of tesserae with full colour (with red substituted for gold). In the spaces where tesserae have fallen off and exposed the bedding mortar, one can see a coloured base coat that serves as a preparatory work for the *musivarius* when laying tesserae.

The use of these frescoes painted: the *campitura*, while often denied in the past, is documented in ancient treatises and codices (Figure 4.2.22).[7]

The setting bed for figures and depiction was lied in *giornate*[8] while the one intended for the gold background and other geometrical decorations was applied in *pontate*. Since the working surfaces were very long, we can also see vertical seams between each *pontata*. In some cases, the *pontate* are masked by compositional schemes such as buildings' architraves on the north wall, which are found near the horizontal seam between two *pontate*.

Figure 4.2.20 North wall, detail of the "*battitura dei fili*".

Figure 4.2.21 North wall, the thickness of the setting bed.

Figure 4.2.22 Transept north: the frescoes painted *campitura* on the setting bed.

Figure 4.2.23 Raking light photography. First angel in the north wall.

Figure 4.2.24 Raking light photography. Foot of one of the apostles in the Ascension scene.

It is clear, from the observations made with raking light,[9] that the first to be executed were the figures and more complex representations, followed by the backgrounds and finally the frames, sometimes even by cutting figurative details such as the apostles' feet in the scene in the north transept. A separate execution was reserved for the faces, hands, and feet of the angels on the north wall of the nave, although they were made almost in the same time as the other figures and sometimes even after the garments (Figure 4.2.23).

In the transept, on the other hand, a master or probably a mosaicist specialised in these representations executed the figures' faces, hands, and feet first, before the rest of the figure (Figure 4.2.24).

This aspect of the mosaics' execution distinguishes the team that worked on the figures in the nave from the team that worked on the figures in the transept.

4.2.3 Tesserae and mosaic texture

The mosaic is made of glass tesserae, gold tesserae, silver tesserae and stone tesserae. One special feature that makes the mosaic even more precious is the presence of mother-of-pearl.

The mother-of-pearl is used in many different shapes and sizes, especially in decorative elements, to embellish the aureoles of Christ and the Virgin and to adorn fabrics and curtains (Figure 4.2.25, 4.2.26).

It is also used in an original way to make some figurative details such as the glass ampulla on the altar in the depiction of the Council of Serdica (Figure 4.2.27).

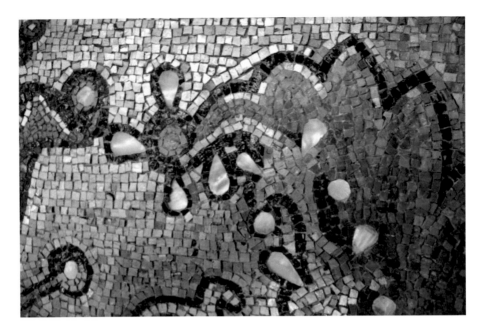

Figure 4.2.25 Mother-of-pearl tesserae.

Figure 4.2.26 Mother-of-pearl tesserae.

Figure 4.2.27 Mother-of-pearl tesserae.

Gold tesserae take on different shades of yellow or gold, depending on the kind of gold leaf used. The glass base, like the *cartellina*,[10] has an amber colour with tones ranging from transparent to dark amber.

Gold tesserae are used mostly in the execution of the backgrounds. Different tones of gold leaf are mixed together to obtain a greater effect. Tesserae are inserted almost completely into the setting bed. Tesserae arranged in parallel lines filled the gold backgrounds, creating a very orderly mosaic texture (Figure 4.2.28).

On some parts, such as backgrounds behind the angels, tesserae are set at a 45° angle to reflect more light when seen from below (Figure 4.2.29).

Tesserae are set in the same way in the aureoles of Christ and Saint Peter in the scene depicting their entry into Jerusalem, as well as in the gemmed cross on the north wall of the nave. Gold tesserae are even set upside down in some specific decorative solutions (Figure 4.2.30).

Figure 4.2.28 The gold background.

Figure 4.2.29 The gold background behind the angels.

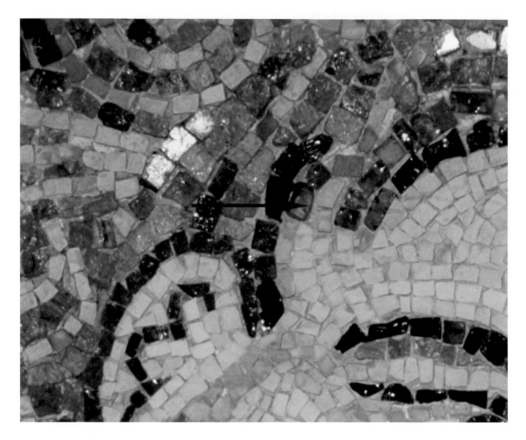

Figure 4.2.30 The red arrow indicates a golden tesserae upside down.

Silver tesserae also have a transparent glass support surface; support and *cartellina* have the same composition. The metallic foil is made of silver with traces of copper.[11]

Silver tesserae cut into rectangular and round shapes and are used to obtain different lighting and decorative effects (Figure 4.2.31).

Use of silver tesserae, though not as extensive as that of the gold tesserae, is nevertheless very frequent. Silver is employed extensively to embellish some specific decorative elements or to create lighting effects such as the rays of light emanating from the cross on the dome of the church in the Council of Antioch (Figure 4.2.32). It is also used for the angels' wings in the nave and in the scenes of the provincial councils on the north wall.

On the south wall of the nave, silver tesserae are employed as background of the inscriptions for the Councils of Ephesus and Chalcedon (Figure 4.2.33).

In the scene of the Transfiguration in the transept, the beam of light that strikes Saint Peter and the bottom frame are made with silver tesserae (Figure 4.2.34). In addition to gold tesserae, silver tesserae create highlights on garments in the figurative depictions. In the scene depicting the entry into Jerusalem, silver tesserae embellish Christ's tunic and make a contrast with his himation made of gold tesserae.

There are three kinds of stone tesserae. For colours ranging from white to grey to light blue most used is the Proconnesian marble, a medium- to coarse-grained white marble with grey streaks (Figure 4.2.35).

Figure 4.2.31 Round silver tesserae.

Figure 4.2.32 North wall. Silver tesserae in the dome of the church in the council of Antioch.

This kind of stone tesserae is extensively employed to create specific shades of colour, especially grey and very light blue, for the garments on the angels in the nave and on the apostles in the transept.

Throughout the mosaic, the second kind of stone employed is a white limestone. The tesserae have a compact and opaque appearance, an irregular shape and a rough surface (Figure 4.2.36).

Figure 4.2.33 The inscriptions of the council of Chalcedon.

Figure 4.2.34 South transept. Silver tesserae.

Figure 4.2.35 North wall. *Proconnesium* marble tesserae.

Figure 4.2.36 North transept. White stone tesserae.

Figure 4.2.37 The third angel's left foot, limestone stone tesserae.

The third kind of stone used is also a compact limestone with colour that ranges from ivory to brick red. This kind of tesserae can be found throughout the mosaic, especially for skin tones. Tesserae are cut smaller for faces, hands and feet than other areas of the mosaic (measuring under 1 centimetre in the nave and 5 to 7 millimetres in the transept) and are used in the natural stones' many colour shades: ivory, pale yellow, ochre, pink and bright red. The tesserae have a compact and opaque appearance, an irregular shape, and a rough surface (Figure 4.2.37).

Angels' faces on the north wall are filled in with white-ivory stone tesserae, while the cheeks and external sides of the forehead are marked by pink tesserae. The yellow ochre shades of this stone are used for shading the figures' skin and shadows.

Many different colours of stone tesserae, ranging from white to dark brown, are used for the faces of Christ's ancestors in the bottom register of the first fragment on the south wall. In this case, the tesserae are employed to make curls of hair and beards, which are always outlined in black tesserae, and to fill in the haloes (Figure 4.2.38).

This kind of stone tesserae fills in the faces of the figures in the transept, while pink tesserae mark their cheeks, the shadows on their necks and the lines on their foreheads. Rows of dark-red tesserae define the contours of the faces, hands and feet.

Glass tesserae are predominant in this mosaic and are used in a variety of colours and hues.

There are blue-glass tesserae in hues ranging from dark blue to cerulean blue (Figure 4.2.39).

There are two different varieties of green tesserae; one is translucent with a dark-green hue (Figure 4.2.40), while the other is opaque with tones ranging from dark green to yellow and is used to depict meadows in many different scenes (Figure 4.2.41).

Figure 4.2.38 Christ's ancestors in the bottom register of the first fragment on the south wall.

Figure 4.2.39 Blue glass tesserae.

Figure 4.2.40 Translucent green tesserae.

Figure 4.2.41 Opaque green tesserae.

Red tesserae are opaque, coloured with copper and iron, and have a deep-red hue. They are always used for the outer contours of faces marking the contours of nose and lips (Figure 4.2.42).

Black colour is obtained using transparent glass tesserae in a shade of very dark yellow-green, dark enough to appear black. These tesserae outline figures, aureoles, and decorative details such as frame, capitals, columns and leaves.

Figure 4.2.42 Opaque red tesserae.

They are employed to draw locks of hair and inner contours of faces and to mark eyebrows and eyes. A single transparent black tessera is always used for the iris.

Transparent glass tesserae in shades ranging from amber to dark green often define curls of hair of the angels in the nave.

Translucent sand-coloured and grey glass tesserae filled with air bubbles enrich the colour palette (Figure 4.2.43).

The image is thoroughly depicted on a rich gold background. Texture[12] is always very orderly, and while the tesserae may vary in size, especially in the depiction of skin tones where smaller tesserae are used, the shape of the tesserae is quite regular. All tesserae are laid flat except for those in the gold backgrounds of the angels on the north wall of the nave, where the tesserae are laid at a 45° angle. The gold backgrounds show very accurate parallel lines, and the same setting is employed in each subsequent *pontata*, showing the care that was taken during the execution of the mosaic (Figure 4.2.44).

While there are no differences in style between the mosaic frame decoration in the nave and the one in the transept,[13] except for those created by the logical rotation of the many teams needed for the execution of such a large and demanding project, we can, however, observe some differences in execution technique.

The tesserae used in the nave are larger, having an area of about 1.3 square centimetres, whereas those employed in the transept have an area between 5 and 7 square millimetres.

The stone tesserae used to depict skin tones are even smaller. In the nave, these tesserae have an area of about 0.5 square centimetres, while in the transept they are just over half that size (Figure 4.2.45).

Figure 4.2.43 Black transparent glass tesserae.

Figure 4.2.44 The very accurate parallel lines of the gold background.

Figure 4.2.45 The different size of the tesserae in the figures of the nave, angel and ancestor of the transept.

Figure 4.2.46 The different tessellation in the nave (on left) and in the transept (on the right).

Tesserae in the transept and nave are of the same colour and material, but the tessellation in the nave is less compact and the mortar between the interstices is visible (Figure 4.2.46).

Analysing the use of tesserae and the texture scheme in the faces of the figures in the nave and comparing them to those in the transept reveals clear compositional differences. The faces of the angels and ancestors are more hieratic, shadows are less marked and colour transitions are more subdued.

The inner outlines of the angels' faces are made through multiple rows of transparent glass tesserae in an amber colour that fades to light green, according to its volume. These rows slowly fade into the white colour of the face thanks to a series of ochre yellow stone tesserae laid in parallel rows. Volume of the forehead is masterfully suggested by a triangular shadow made in pink stone tesserae. Darker shades start from the outer edge and converge into two rows of ochre tesserae. A row of opaque red glass tesserae marks the outer contours of the face and nose and draws the lips and ears. The inner shadow of the nose and the shading below the mouth and eyelids are made with rows of transparent glass tesserae. As previously indicated, transparent black glass tesserae mark the contours of hair, eyebrows and perimeters of the eyes. The tesserae used to draw eyebrows are larger towards the tip of the nose, becoming thinner towards the temple.

Figure 4.2.47 The second angel in the north wall.

Hair is rendered through multiple shades of transparent amber-coloured glass tesserae. Hair on top of the figures' heads is especially luminous thanks to the use of gold tesserae inserted face down (Figure 4.2.47).

Two rows of transparent black tesserae laid in concentric circles outline the haloes. Of the latter, those of the first and last angels are filled in with gold tesserae, which, on the right side, are laid in concentric rows around the outer contour of the face, and on the left side, are laid in parallel rows. Nevertheless, haloes of the other angels are filled in with tesserae laid in concentric rows.

Comparing the angels in the nave to the two angels beside the Virgin in the scene of the *Ascension* reveals the differences in technique between the two or more teams of mosaicists who worked simultaneously on the church's mosaic surfaces (Figure 4.2.48).

The faces of the angels beside the Virgin are made entirely of stone tesserae, which, as previously stated, are smaller than the one used for the faces of the angels in the nave. The tesserae lay closer together. Faces stand out against haloes made of concentric rows of gold tesserae and are so detailed to look like being drawn on. Volume is created by the contrast between the light-coloured tesserae used for the faces and the rows of dark tesserae used to outline them. Outer contours of the face and nose are rendered with a single row of red glass tesserae used also for the lips and folds in the neck. Eyebrows are made with transparent black glass tesserae that are all the same size and laid out in arches. Hair locks, outlined in transparent black glass tesserae, are short and rigid; they do not fall gently to the neck like the hair on the angels in the nave. The wings of the angels in the nave, while depicted in the same position, are closer to the figure, and the gold and silver highlights are created with soft lines that curve slightly downwards. Coloured backgrounds are laid in the same pattern. The wings of the angels in the transept are depicted in a more rigid and schematic way, so that the tip of wings that pass behind the haloes

Figure 4.2.48   The angel in the Ascension, on the left of the Virgin.

are not depicted at all. The same considerations can be made for the figures of the apostles and of Christ's ancestors depicted on the south wall of the central nave.

In conclusion, the studying of the executive techniques, of the typology of the materials and of their characteristics brings us back to the thesis put in the preamble: the mosaic decoration of the basilica was realised in the same arc of time by several teams of mosaicists who worked using different techniques from their school or tradition.

References

[1] Alberti, L., Bourguignon, E., Carbonara, E., Roby, T. & Segura Escobar, J. (2013) *Illustrated Glossary: Technician Training for the Maintenance of In Situ Mosaics*. Los Angeles; Getty Conservation Institute.

[2] Bianchetti, P., Santopadre, P., Profilo, B. & Verità, M. (1993) Il restauro del mosaico di Santo Stefano Rotondo a Roma: studio dei materiali. *Arte Medievale*, II serie, 7(1), 211–218.

[3] Brandi, C. (1956) Nota sulle tecniche dei mosaici parietali in relazione al restauro e lale datazioni. *Bollettino Istituto Centrale del Restauro*, 25–26, 3–9.

[4] Currzi, G. (1993) La decorazione musiva della Basilica dei SS Nereo e Achilleo in Roma: materiali e ipotesi, Encicòopedi Italiana, *Arte Medievale*, II serie, 7(2), 21–45.

[5] Farneti, M. (1993) *Glossrio tecnnico-storico del mosaico*. Ravenna.

[6] Fisher, P. (1971) *Mosaic: History and Technique*. Thames and Hudson, London.

[7] Haswell Melletin, J. (1973) *The Thames and Hudson Manual of Mosaic*. Thames and Hudson, London.

[8] Merrifield, M. (1967) *Original Treatises on the Arts of Painting*, Volume 1, cap. III the Getty Institute Conservation. New York, pp. L–LII.

[9] Mora, P., Mora, L. & Philippot, P. (1984) *Conservation of Wall Paintings*. Butterworths, London.

[10] Wilkinson, J. (1977) *Jerusalem Pilgrims before the Crusaders*. Jerusalem.

For high-resolution tables of the following five tables, please check the Additional Resources tab at https://www.crcpress.com/9781138488991

Table 4.3.1 Graphic survey tessera by tessera. Mosaics in the northeast wall.

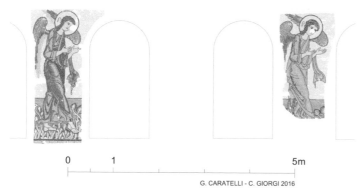

Table 4.3.2 Graphic survey tessera by tessera. Mosaics in the northwest wall.

Table 4.3.3 Graphic survey tessera by tessera. The South wall mosaics.

Table 4.3.4 Graphic survey tessera by tessera. The North transept mosaics.

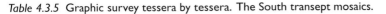

Table 4.3.5 Graphic survey tessera by tessera. The South transept mosaics.

4.3   Mosaics materials analysis

M. Verità, L. Lazzarini and P. Santopadre

The tesserae of the wall mosaics are made of glass or stone set on a support consisting of two layers of mortar. The first layer is applied directly on the wall; the second is the layer onto which the tesserae are set. The analytical investigation of these materials is presented and discussed.

4.3.1   Mortar analysis (P.S.)

The support of the tesserae of the wall mosaics of the Nativity Church, Bethlehem, consists of two layers of mortar. The first layer is applied directly on the wall; the second is the layer onto which the tesserae are set following the preparatory painting below. This chromatic pattern provided a useful guide to the mosaicist for the setting of the tesserae and helped to make the interstices less visible (James, 2017); it is often revealed again when the tesserae fall away (Figure 4.3.1).

Three samples of mortar, some cm in length, were collected to be analysed. One sample of the first layer (sample 1) and one sample of the second layer (sample 2) were collected from the northern bema decoration; one sample of the second layer (sample 3) was collected from the nave, southern wall, near the face of one of Christ's ancestors.

The samples were first observed under the optical microscope in reflected light; small fragments were then cut and ground to a thin powder to be analysed by X-ray diffraction (XRD). Part of the remaining mortar of each sample was glued to a glass support with epoxy resin and thinned by abrasion with silicon carbide powders down to a thickness of 30 micrometers to be studied under the optical polarised microscope in reflected/transmitted light. Organic substances were searched by thermogravimetric analysis (TGA)[14] and by Fourier transform infrared spectroscopy (FTIR) (Santopadre *et al.*, 2017).

The analyses showed that the first mortar layer is made of lime with clusters of micritic and sparitic limestones. Vegetal fibres were also added as a filler and strengthening material (Figures 4.3.2 and 4.3.3). XRD analysis revealed the presence of calcite as the main

Figure 4.3.1 North bema, mortar bed area (layer 2) painted in red where some gold tesserae are missing.

Figure 4.3.2 Sample 1, vegetal fibres embedded in the mortar.

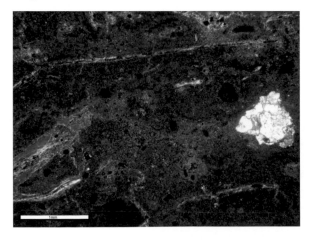

Figure 4.3.3 Sample 1, polarised-transmitted light micrograph of the thin section (crossed nicols). Several vegetal fibres can be observed and, on the right, a fragment of sparitic limestone.

component and traces of quartz and feldspars. FTIR analysis shows the peaks pattern typical of calcite. About 2% of organic compounds were detected by TGA on a sample from which the vegetal fibres were previously removed.

The examination of the fragments and sections of the second layer (setting bed), revealed that the mortar consists of carbonatic micritic matrix with aggregates (Figure 4.3.4) containing more abundant clusters of micritic and sparitic limestones as compared to the first layer. Few reddish lithic fragments were also present, together with traces of vegetal fibres that were also observed in the mortar fragments (Figure 4.3.5). The XRD and FTIR analyses detected the typical signals of calcite and the TGA indicated the presence of about 2% organic matter, as in sample 1. This small amount could be attributed to residual vegetal fibres not completely removed from the analysed sample.

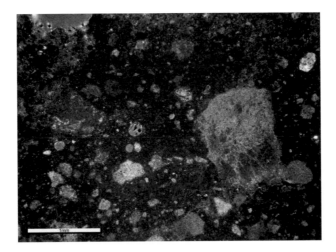

Figure 4.3.4 Sample 3, polarised-transmitted light micrograph of a thin section (crossed nicols) showing several micritic limestone fragments of different sizes.

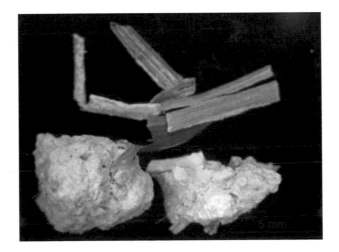

Figure 4.3.5 Sample 3, optical micrograph in reflected light with several vegetal fibres.

The analyses allow us to conclude that mortars of layer one and two were prepared with lime and limestone clusters. Vegetal fibres were also added, probably with the aim to minimize the shrinkage of the mortar during drying, control the setting time, improve the mechanical strength and obtain light mortars.

4.3.2   Stone tesserae analysis (L.L.)

The mosaics of the Nativity Church of Bethlehem are mainly made of glass tesserae, but white, flesh-tone (i.e. opaque pinkish) and dark-grey tesserae are exclusively made of stone. Stone tesserae are present in all the wall mosaicated areas of the church. From the mosaic decoration of the bema northern surface (Figure 4.3.6), one tessera for each of these three colours was sampled for analysis.

The tesserae were studied by two analytical techniques: X-ray diffraction to identify the main minerals present and examination of thin sections by transmitted polarised light to reach a first petrographic classification of the stones. Small fragments cut from each tessera were ground to a thin powder to be analysed by X-ray diffraction. The remaining slice of each tessera was glued to a glass support by epoxy resin and thinned by abrasion with silicon carbide powders down to 30 micrometers to be studied under optical polarised microscopy in transmitted light. The results are summarised below.

Grey tessera (BEn-St1) – The X-ray diffraction analysis revealed the presence of calcite (calcium carbonate, main mineral) and of traces of quartz (silicon oxide). The study of the thin section has shown the typical mortar fabric of a metamorphic marble (Figure 4.3.7) formed by calcite crystals with embayed/sutured boundaries having an MGS (maximum grain size) of 2.02 mm. Small amounts of the following accessory minerals were also observed:

- carbonaceous matter/graphite (amorphous/crystalline C), responsible for the grey colour of the marble);
- traces of quartz, K-mica, apatite (calcium phosphate), titanite (CA-Ti silicate), pyrite (iron sulphide), hematite (anhydrous iron oxide).

Figure 4.3.6 Detail of the mosaic decoration of the bema, northern wall, where the stone tesserae were sampled for analysis.

Figure 4.3.7 Micrograph of the thin section of the dark-grey tessera Ben-St1 of Proconnesian marble observed in transmitted polarised light (crossed nicol, long side: 2.35 mm).

Based on the above minero-petrographic features, and by direct comparison with the database of the most important crystalline marbles used in antiquity (Antonelli and Lazzarini, 2015), it can be concluded that this tessera has been cut from a slab of Proconnesian marble (from the ancient island named Proconnesus, now Marmara Adasi, Turkey, in the Marmara sea).

White tessera (BEn-St2) – The X-ray diffraction analysis has given all the peaks of calcite (main mineral) and a few peaks of quartz, present in traces. The microscopic study of the thin section has shown a microcrystalline fabric composed by micrite (calcite crystals less than 4 micrometers in diameter), with microsparitic areas of the larger grain size (Figure 4.3.8). Small amounts of dispersed (or concentrated in globular aggregates) yellow-brownish ochraceous particles (limonite + hematite) are present in areas. This stone may be classified petrographically as a micritic limestone (Folk, 1959).

Flesh-tone tessera (BEn-St3) – The X-ray diffraction analysis has indicated the main presence of magnesite (magnesium carbonate), with small amounts of dolomite (calcium-magnesium carbonate) and of traces of quartz. The microscopic examination of the thin section has shown a crystalline fabric formed by microsparitic crystals of magnesite and dolomite with abundant small red earthy aggregates (or single particles) of hematite (Figure 4.3.9) and rare individuals of detritic quartz. The stone may be classified as an impure magnesite.

While the grey tessera of Proconnesian marble was very likely produced from re-used slabs of what was the most commonly used marble in antiquity largely exported to the whole Mediterranean area in the Roman-Byzantine periods (Lazzarini, 2015), it is probable that the white and the flesh-tone tesserae are from local geological formations of limestones and dolomias (respectively). The flesh-tone magnesite-tessera was likely obtained from red levels sometimes common in dolomias.

Figure 4.3.8 Micrograph of the section of the white limestone tessera BEn-St2 observed in transmitted, polarised light (crossed nicol, long side: 0.93 mm).

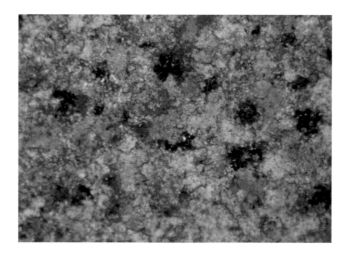

Figure 4.3.9 Micrograph of the section of the flashed tone tessera BEn-St3 mainly composed of magnesite with red hematite particles observed in transmitted, polarised light (crossed nicol; long side: 0.93 mm).

4.3.3   Glass tesserae analysis (M.V.)

Before sampling, a close examination was performed in the mosaic areas still surviving in the basilica; the sampling areas and labels of the tesserae are reported in Table 4.3.1. The mosaic in the small apse of the cave was not sampled.

The examination was made before cleaning of the mosaics. The layer of dirt firmly adhering to the rough surface of these tesserae hampered to precisely define their colour (Figure 4.3.10).

Table 4.3.1 Sampling area and labels of the analysed glass tesserae

Tesserae	Sampling area
NAa	Nave, northern wall Angels
NAn	Nave, northern wall Councils
NAs	Nave, southern wall Councils
TRn	Transept, northern-eastern wall
TRs	Transept, southern-eastern wall
BEn	Bema, decoration northern wall
BEs	Bema, inscription southern wall
NX	Loose tesserae from the materials accumulated over the narthex vaults

Figure 4.3.10 Grainy semi-opaque glass tesserae before cleaning.

The examination of the mosaics before cleaning revealed that they are made mainly of glass tesserae with many mother-of-pearl inserts. White, flesh-tone and dark-grey colours are exclusively made of stone tesserae (Figure 4.3.11).

Four typologies of glass tesserae were identified, according to their aspect:

- gold- and silver-leaf tesserae;
- Semi-opaque tesserae, coloured in blue, anise, brown, purple, light grey and green with a rough surface difficult to clean;
- translucent to opaque, bright tesserae with smooth, easy-to-clean surface, in a wide range of colours from turquoise to green and yellow, including several intermediate hues;
- red opaque and black translucent tesserae, also bright in aspect and with a smooth surface.

Figure 4.3.11 Flesh-tone and white tesserae made of stone in the face of a Christ ancestor (nave, southern wall).

These four typologies of glass tesserae are present in all the mosaicated areas of the basilica. A minimum number of tesserae was collected for analysis in order to limit even slight damage to the mosaics. As a whole, 20 glass tesserae were sampled from the five mosaicated areas with the help of the restorer's staff.

Moreover, 15 tesserae were selected for the analyses among a hundred glass tesserae recovered from the materials accumulated over the narthex vaults. It is not clear when this material was deposited and where it came from. Because the colours and aspect of these tesserae are similar to those of the mosaics, it seemed appropriate to analyse a number of them in order to examine some peculiar aspects which could not be investigated on the church mosaics except by taking an unacceptable amount of samples, that is the chromatic range of the gold tesserae and the turquoise to green and yellow hues.

Small fragments (less than 3 mm in size) were taken from each coloured tessera. In the case of metal (gold and silver) leaf tesserae, a thin diamond wheel was used to cut a section 1–2 mm thick including the protective glass layer (cartellina), the metal leaf and the support glass, so that the three materials could be analysed separately. After sampling, the tesserae were reinserted in the mosaics by the restorers.

Samples were embedded in acrylic resin, and cross sections were ground and polished down to 1 μm diamond pastes. After examination under the optical microscope the samples were carbon coated under vacuum. Quantitative chemical analysis of glass and metal leaves and identification of the opacifiers and pigments were performed at the SEM (Philips XL30) by energy dispersive X-ray microanalysis (EDAX Ametek). SEM observation was made in backscattered mode to allow areas with different chemical composition in the tesserae to be identified (darker grey indicates areas where lighter elements prevail). The blue tesserae were also analysed by an energy dispersive X-ray microfluorescence spectrometer (μEDXRF) Bruker M4 Tornado to determine concentrations of trace elements with a better sensitivity.

4.3.3.1 Results

The quantitative chemical composition of the tesserae is reported in weight percent of the oxides in Table 4.3.2 (coloured tesserae) and Table 4.3.3 (metal leaf tesserae).

4.3.3.1.1 GLASS COMPOSITION

The glass of the tesserae is composed mainly of silica (SiO_2), sodium (Na_2O) and calcium (CaO) oxides. By reporting in diagrams the contents of potassium, magnesium (Figure 4.3.12) and phosphorus, two compositional groups were identified: natron glass and soda plant ash glass. A change in glassmaking technology from natron glass to plant ash glass occurred in the Near East (Byzantine and Islamic glass) and in Italy around the late eighth to early ninth century. It was a long transition; as far as is known, soda plant ash glass replaced natron glass completely only since the end of the 12th early 13th century (Phelps *et al.*, 2016; Verità, 2017).

The tesserae having potassium and magnesium oxides contents below 1.5% and a phosphorus content below 0.2% can be classified as natron glass.

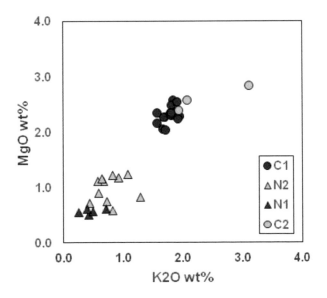

Figure 4.3.12 Potassium versus magnesium concentrations. The compositional groups are indicated with different symbols: triangles, natron glass; circles, soda plant ash glass.

Table 4.3.2 Quantitative chemical composition in wt% of the oxides of the coloured glass tesserae. Opacifiers: Qu, quartz; CaP, bone ash; Sn, tin oxide; Sn, PbSn, tin oxide and lead stannate; PbSn, lead stannate; Cu, metallic copper; tr, transparent. n.d., not detected.

Sample	Colour	Opacifier	Glass Type	SiO_2	Al_2O_3	Na_2O	K_2O	CaO	MgO	SO_3	P_2O_5	Cl	TiO_2	Fe_2O_3	MnO	CuO	CoO	ZnO	SnO_2	PbO
NAs-B4	Blue	Qu	CI	67.8	2.25	12.3	1.87	8.60	2.20	0.15	0.20	0.58	0.15	2.43	0.90	0.10	0.13	n.d.	n.d.	0.32
TRs-B17	Blue	Qu	CI	68.5	2.05	12.4	1.65	8.70	2.20	0.17	0.20	0.50	0.10	2.35	0.70	0.10	0.11	n.d.	n.d.	0.30
NAa-B10	blue dark	Qu	CI	65.4	2.30	13.8	1.82	9.20	2.45	0.19	0.22	0.53	0.12	2.80	0.70	0.08	0.12	n.d.	n.d.	0.35
NAn-B15	blue light	Qu	CI	71.6	2.05	11.7	1.60	7.60	2.00	0.06	0.20	0.43	0.10	1.65	0.50	0.12	0.12	n.d.	n.d.	0.30
BEn-B20	blue light	Qu	CI	71.6	2.00	12.0	1.65	7.40	2.00	0.07	0.20	0.53	0.10	1.40	0.75	0.07	0.09	n.d.	n.d.	0.15
NAn-CI2	Anise	Qu	CI	68.5	1.90	12.5	1.50	8.40	2.26	0.25	0.20	0.65	0.12	0.75	0.80	2.20	n.d.	n.d.	n.d.	n.d.
NAn-CI4	Anise	Qu	CI	69.0	2.15	12.0	1.50	8.10	2.07	0.30	0.30	0.57	0.10	0.87	0.82	2.20	n.d.	n.d.	n.d.	n.d.
NAn-C2	Grey	CaP	N2	62.7	2.30	16.8	0.80	10.5	1.20	0.30	1.80	0.75	0.22	1.02	1.57	n.d.	n.d.	n.d.	n.d.	n.d.
TRs-CI6	Aqua	CaP	N2	65.0	2.90	14.4	1.25	10.3	0.82	0.20	1.20	0.67	0.15	0.62	0.36	1.60	n.d.	n.d.	n.d.	0.45
NX-D29	Turquoise t transl.	Sn	N2	67.2	1.90	17.6	1.05	6.70	1.22	0.36	0.27	0.85	0.21	0.60	0.12	1.35	n.d.	n.d.	0.20	0.35
NX-D30	Turquoise t transl.	Sn	NI	70.0	2.68	17.3	0.40	3.00	0.50	0.36	n.d.	1.00	0.25	0.95	0.08	2.50	n.d.	0.15	0.30	0.50
NAn-C13	green dark	Sn; PbSn	N2	59.0	2.10	20.0	0.42	5.92	0.65	n.d.	n.d.	1.10	0.14	0.67	0.10	1.37	n.d.	n.d.	0.90	7.60
NX-D32	green dark	Sn; PbSn	NI	64.8	3.24	16.5	0.43	3.10	0.52	n.d.	n.d.	0.80	0.31	0.98	0.10	1.65	n.d.	0.15	0.60	6.80
NAs-C5	green light	Sn; PbSn	N2	58.5	2.67	13.0	0.70	8.00	0.50	n.d.	n.d.	n.d.	0.10	0.43	0.08	1.40	n.d.	n.d.	1.60	13.0
NX-D33	green light	Sn; PbSn	NI	65.0	1.70	16.5	0.22	3.60	0.50	n.d.	n.d.	0.80	0.20	0.70	0.10	0.93	0.10	0.10	1.20	8.50
NX-D34	yellow green	PbSn	NI	57.5	2.25	15.8	0.57	2.65	0.50	n.d.	n.d.	n.d.	0.20	0.72	0.10	0.20	n.d.	n.d.	2.00	17.5
NAn-CII	Yellow	PbSn	N2	44.8	2.22	10.7	0.50	6.20	0.50	n.d.	n.d.	n.d.	0.07	0.52		0.35	n.d.	n.d.	5.10	29.0
NX-D35	Yellow	PbSn	NI	50.5	1.75	13.0	0.27	3.50	0.44	n.d.	n.d.	0.19		0.66	0.35	0.06	n.d.	n.d.	4.30	25.0
TRs-AI8	Red bright	Cu	CI	64.0	2.00	14.3	1.70	8.52	2.35	0.15	0.15	0.52	0.18	2.69	0.70	1.70	n.d.	0.25	n.d.	0.80
NAs-A6	Red bright	Cu	CI	64.4	2.35	13.6	1.60	8.20	2.15	0.45	0.70	0.62	0.20	2.85	0.95	1.35	n.d.	0.10	n.d.	0.45
NAs-A7	Red brown	Cu	CI	65.0	2.10	12.5	1.72	8.80	2.43	0.40	0.25	0.53	0.18	2.88	1.00	1.60	n.d.	0.15	n.d.	0.45
BEs-C9	black	tr	CI	69.0	1.55	11.4	1.85	9.18	2.30	0.10	0.20	0.45	0.10	3.05	0.57	0.25	n.d.	n.d.	n.d.	n.d.
NX-D28	Black	tr	CI	66.0	2.42	12.7	1.80	8.90	2.25	0.10	0.20	0.63	0.10	3.70	0.95	0.20	n.d.	n.d.	n.d.	n.d.

Table 4.3.3 Quantitative chemical composition in wt% of the oxides of the glass of the metal leaf tesserae and in wt% of the element of the metal leaves. n.d., not detected.

Sample	Glass colour	Glass Type	Glass												Metal leaf		
			SiO_2	Al_2O_3	Na_2O	K_2O	CaO	MgO	SO_3	P_2O_5	Cl	TiO_2	Fe_2O_3	MnO	Au	Ag	Cu
NX-OE21-Au(s)	colourless	N2	65.0	2.50	19.3	0.62	7.45	1.15	0.42	n.d.	1.05	0.20	0.83	1.44			
NX-OE22-Au(sc)	colourless	N2	67.0	2.35	19.0	0.57	6.70	0.90	0.32	n.d.	0.87	0.13	0.80	1.35	98.0	2.0	n.d.
NX-OE27-Au(c)	Brown layered	N2	65.4	3.30	14.2	0.88	10.2	1.15	0.12	n.d.	0.63	0.15	0.75	3.20	88.0	12.0	n.d.
NX-OE27-Au(s)	olive green	N2	64.0	3.40	14.0	0.84	11.2	1.20	0.10	n.d.	0.65	0.20	1.10	3.30			
NAn-AI-Au(sc)	colourless	N2	65.3	2.50	19.0	0.55	7.75	1.10	0.30	n.d.	0.92	0.13	0.95	1.50	92.0	8.0	n.d.
NAn-A8-Ag(c)	colourless	N2	66.0	2.70	17.4	0.65	8.60	1.10	0.30	n.d.	0.85	0.15	0.85	1.40	n.d.	99.3	0.7
NAn-A8-Ag(s)	colourless	N2	65.2	2.70	17.4	0.62	9.00	1.15	0.25	n.d.	0.85	0.15	0.95	1.70			
NX-OE23-Au(sc)	colourless	C1	68.0	2.60	13.4	1.77	8.55	2.27	0.23	0.23	0.77	0.15	0.58	1.42	98.0	2.0	n.d.
NX-OE24-Au(c)	colourless	C1	68.0	2.60	13.5	1.75	8.55	2.30	0.20	0.20	0.70	0.18	0.58	1.40	94.0	6.0	n.d.
NX-OE24-Au(s)	Brown layered	C1	68.0	2.60	13.3	1.80	8.58	2.30	0.20	0.20	0.75	0.15	0.58	1.58			
NX-OE25-Au(c)	colourless	C1	68.0	2.65	13.6	1.73	8.50	2.31	0.20	0.20	0.75	0.16	0.62	1.32	90.0	10.0	n.d.
NX-OE25-Au(s)	Brown layered	C1	67.6	2.60	13.6	1.75	8.55	2.27	0.18	0.20	0.75	0.20	0.60	1.72			
NX-OE26-Au(c)	colourless	C1	68.0	2.60	13.8	1.77	8.50	2.32	0.10	0.23	0.75	0.15	0.62	1.15	91.5	8.5	n.d.
NX-OE26-Au(s)	Brown layered	C1	67.8	2.60	13.4	1.87	8.70	2.20	0.10	0.25	0.70	0.10	0.53	1.75			
BEs-A3-Au(sc)	colourless	C2	70.5	1.70	12.7	2.05	8.15	2.55	0.25	0.22	0.67	0.12	0.40	0.70	96.5	3.5	n.d.
TRn-Al9-Au(c)	colourless	C2	66.5	1.30	13.9	3.05	9.30	2.80	0.15	0.20	0.92	0.15	0.57	1.25	94.0	6.0	n.d.
TRn-Al9-Au(s)	Brown layered	C2	66.5	1.30	13.9	3.00	9.15	2.90	0.18	0.22	0.90	0.10	0.38	1.50			

Natron is a sodium carbonate mineral associated to low amounts of other salts such as chlorides and sulphates extracted in Egypt, used to produce glass. It was mixed and melted together (glass batch) with a silica-lime sand in which quartz and calcium carbonate were present in suitable ratios to make glass. Sands with such properties were quarried in a few sites, identified up to now in the Levantine area and in Egypt (Freestone, 2005).

At least until the ninth century the glass production occurred in two stages. The melting of the batch was performed in large tank furnaces of capacity up to 40 tons of glass. Similar furnaces were uncovered up today in Israel and Lebanon of today and in Egypt, near the raw materials sources. When the glass was ready, the furnace was demolished and the recovered glass blocks (called raw or primary glass) were transported to secondary workshops to be remelted in small furnaces, coloured, opacified and manufactured in the form of glass cakes, from which the tesserae were subsequently cut. For this reason the chemical composition of the glass of the tesserae is not specific to the workshop producing the glass cakes but rather to the primary melting furnace. This fact limits the information supplied by the analysis of the glass composition.

By reporting in a diagram the contents of calcium and potassium oxides, the natron-type tesserae can be divided into two groups (Figure 4.3.13).

Five natron glass tesserae of turquoise to yellow colours from the narthex (N1 group, dark triangles) are characterised by low calcium (CaO 3–5%) and high titanium (TiO2 0.23–0.35%) concentrations. Tesserae made in the same range of colour and with a quite similar composition were found in the site of Qusayr'Amra (Jordan), an early Islamic desert castle built during the first half of the eighth century (Verità and Santopadre, 2017). Similar compositions were also found in glass artworks from the Omayyade period (second half of seventh-mid eighth century) for which an Egyptian provenance has been hypothesised (Gratuze and Barrandon, 1990).

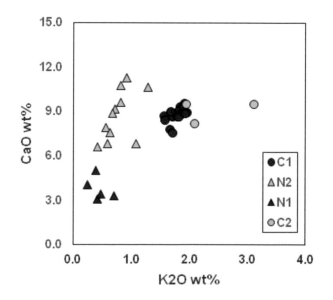

Figure 4.3.13 Potassium vs calcium concentrations. The compositional groups are indicated with symbols as in Figure 4.3.12.

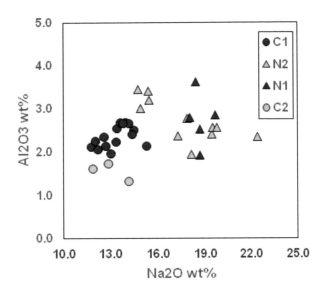

Figure 4.3.14 Sodium vs aluminium concentrations. The compositional groups are indicated with symbols as in Figure 4.3.12.

The remaining 11 natron type tesserae (N2 group, grey triangles in the diagrams) were sampled in all the mosaic areas (northern and southern walls of the nave, transept) and in the narthex. By reporting the sodium and aluminium oxide concentrations in a diagram, N2 type tesserae split into two sub-groups corresponding to two recipes (Figure 4.3.14). The first group includes four coloured, low-sodium (Na2O 14.7–15.4%) and high-aluminium (Al2O3 3.0–3.5%) tesserae; to the second group belong seven coloured and gold-leaf tesserae showing high-sodium (Na2O 17–22.4%) and low-aluminium (Al2O3 1.9–2.8%) concentrations. These glasses were made with less natron and more silica-lime sand in the first group and with less natron and more sand in the second group. As no other significant difference was detected, it is likely that the two groups were melted with the same sand of Levantine origin (Freestone, 2005).

The tesserae of the second group (indicated with C2, three gold tesserae from the transept) are made of soda plant ash glass characterised by higher potassium, magnesium and phosphorous levels and lower sodium concentrations (Na2O 12–14.5%) compared to the natron-type glass tesserae. Ashes of halophytic plants (plants growing in saline and semi-desert areas) rich in soda and calcium carbonate were used as a flux of a silica source (quartz) for the glass production.

Again, the composition of these tesserae allows two sub-groups to be distinguished. Most of the tesserae belong to a group indicated with C1, in which the potassium content (K_2O 1.6–1.9%) is slightly lower than 2%, i.e. the amount conventionally assumed as an indicator of the use of soda plant ash as a flux, being intermediate between natron-type glass and soda plant ash–type glass. Instead, the amount of magnesium (MgO > 2%) and phosphorous (P_2O5 > 0.2%) fall within the limits of the soda plant ash glass type. Similar compositions were found in 11th to 13th century Byzantine mosaic tesserae in Italy (Torcello: Henderson and Roe, 2006; Verità and Zecchin, 2012, and Monreale: Verità and Rapisarda, 2008) and Greece (HosiosLoukas, Freestone *et al.*, 1988; Brill, 2008; Arletti *et al.*, 2010; and Daphni, Arletti *et al.*, 2010). This composition was found also in blown-glass artefacts from Italy (Uboldi and Verità, 2003) and Middle Eastern countries (Henderson *et al.*, 2004). Some authors have interpreted this composition as a "mixed natron–soda plant ash composition" obtained by melting together natron glass and soda ash glass. This is possible, but other interpretations are also plausible (the use of a particular type of plant ash, for instance). To present knowledge, glass with this composition was used only for items manufactured between the 11th and the end of the 12th century.

4.3.3.1.2   GROUND QUARTZ TESSERAE

Translucent tesserae coloured in blue, anise, green, brown, purple and grey have been widely used in the mosaics (Figure 4.3.15). Their surface is rough and uneven, so it strongly retains dirt and makes cleaning difficult.

The five blue and the two anise tesserae analysed are made of soda plant ash glass of C1 type opacified with quartz particles varying in size from a few micrometers to 0.3 mm and with a considerable amount of bubbles, giving the tesserae a characteristic translucent aspect. The sharp-edged quartz grains (SEM micrograph in Figure 4.3.16) demonstrate that ground quartz was added to the molten glass immediately before making the glass cakes from which the tesserae were cut. The dark hues were obtained by adding small amounts of (or without any addition of) quartz to the coloured glass; the light hues by adding large amounts. This technique allowed three or four hues of a certain colour to be obtained. The anise tesserae were coloured with copper (CuO 2.2%) added as a pure oxide to a transparent glass decolourised with manganese (MnO 0.8%).

Blue tesserae sampled one for each area of the surviving mosaics were chosen as a "marker" to identify any difference in the use of glass tesserae. The choice was due to the fact that blue tesserae are abundantly present in all the five areas of the mosaics, they are

Figure 4.3.15 Clothes of an angel made of glass tesserae opacified with ground quartz. Nave northern wall.

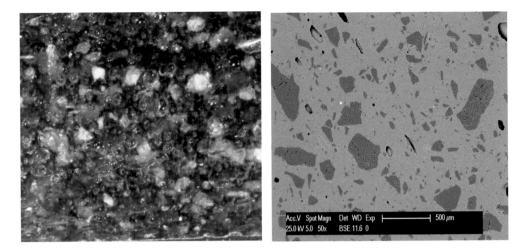

Figure 4.3.16 Optical (left) and SEM micrographs of the polished cross section of the anise tessera NAn-C12, opacified by translucent grains of quartz. The SEM micrograph (right) evidences the sharp edges grains (dark grey).

opacified with the same technique (ground quartz and bubbles) and the identification of the cobalt ore added to colour the tesserae can supply additional information about their provenance and dating. The analyses of the blue tesserae show similar characteristics: the glass is of the soda ash C1 type opacified with ground quartz and bubbles. The blue colour is given by small amounts of cobalt (CoO 0.09%–0.13%) added as a cobalt ore in which iron and small amounts of copper and lead were also present. Nickel, bismuth, arsenic and zinc (fingerprint elements of different types of cobalt ore) were searched for but not found by X-ray fluorescence analysis, which allows these elements to be detected down to 0.01%.

Tesserae with the same characteristics have been found in Byzantine mosaics dated to the 11th to 13th centuries (Torcello and Monreale in Italy, Daphni and Hosios Loukas in Greece and Hierapolis in Turkey). To present knowledge, C1-type glass was used only in the transition period from natron glass to soda ash glass, that is between the 9th and 13th centuries. The opacification with ground quartz has been identified up to now in mosaics dated to not before the ninth century (Neri *et al.*, 2018). In this period, blue tesserae made of C1-type glass opacified with quartz were coloured also with a cobalt ore containing also zinc (Verità & Zecchin, 2012). The use of this cobalt ore was not detected in the Bethlehem blue tesserae. These results allow concluding that all the wall mosaics survived in the Nativity Church have been set up with glass tesserae having compatible characteristics for a production spanning over the period from the 11th to the 13th centuries.

4.3.3.1.3   BONE ASH TESSERAE

Tesserae NAn-C2 (grey) and TRs-C16 (aqua) are heterogeneous due to a great amount of small bubbles and white particles (from 5 micrometers to 0.2 mm) roughly dispersed in natron N2-type glass (Figure 4.3.17). The surface of these tesserae is brighter and less rough than that of the tesserae opacified with quartz because of the smaller size of the bubbles and

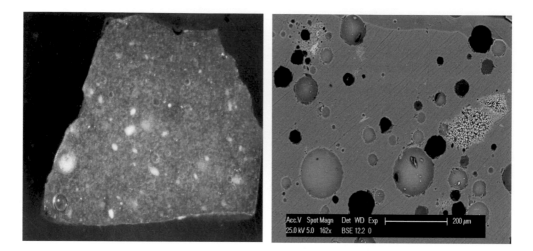

Figure 4.3.17 Optical micrograph (left) and SEM micrograph (right) of the polished section of the aqua tessera TRs-C16 opacified with bone ash.

less spiky particles. The natron-type glass composition of these tesserae shows an unusually high concentration of phosphorous (P_2O_5 1.2–1.8%). SEM micrographs provide evidence of particles, sometimes spongy in aspect (Figure 4.3.17). Probably their porous structure causes gas release and formation of large amounts of small bubbles. Chemical analysis revealed that coarser particles consist mainly of calcium and phosphorous oxides (CaO 55%; P_2O_5 43%, calcium phosphate), whereas in smaller particles sodium is also present (Na_2O up to 18%). These data indicate that they are bone ash particles. The presence of sodium results from the dissolution process of the bone ash in the glass melt (Verità, 2010; Silvestri *et al.*, 2012).

Bone ash–opacified tesserae are peculiar Byzantine production, first identified among the tesserae excavated in a basilica at Petra (Jordan) active between the middle fifth and early eighth centuries (Mahri and Rehren, 2009), then in northern Italy (probably imported from the Levant) in fifth-century mosaics in Ravenna (Verità, 2010) and in Padua (Silvestri *et al.*, 2012) and in the eighth-century archaeological site of Qusayr' Amra, Jordan (Verità and Santopadre, 2017). Similar tesserae were found in other sites of the Levantine area; as far as we know, all these mosaics date to before the end of the ninth century. After that date bone ash was probably replaced by ground quartz as a glass opacifier in the manufacture of mosaic tesserae.

It is therefore likely that the bone ash tesserae were not expressly made for the investigated mosaics of the Nativity Church but were recovered from ancient dismantled mosaics (in any case made after the fifth century) and reused. There are no grounds for assessing whether these dismantled mosaics were from the Nativity Church or other sites.

4.3.3.1.4   TURQUOISE, GREEN, AND YELLOW TESSERAE

Another abundant group of tesserae of the Bethlehem wall mosaics is composed of opaque samples with a smooth and bright surface in a large range of colours from turquoise to green

Figure 4.3.18 A detail of the mosaic with turquoise to green and yellow tesserae; nave northern wall.

and yellow through a series of intermediate hues (Figure 4.3.18). Nine tesserae were selected for analysis and classified as follows: turquoise translucent (NX-D29 and D30), dark green (NAn-C13 and NX-D32), light green (NAs-C5 and NX-D33), yellow-green (NX-D34) and yellow (NAn-C11 and NX-D35). Three tesserae are from the nave mosaics; the others are loose tesserae from the narthex. All these tesserae are of natron-type glass: N2 type the ones from the nave and NX-D29; N1 type the other tesserae from the narthex.

The colourants of this group of tesserae were copper and yellow pigments added to a not (or slightly) decolourised glass (in general, MnO < 0.1%). The turquoise translucent tesserae are made of glass coloured by oxidised copper. The presence of zinc detected in tessera NX-D30 (N1-type glass) and not in tessera NX-D29 (N2-type glass) attests the use of different copper sources: a metallurgical copper-zinc by-product and pure copper oxide, respectively. This pattern, that is present also in the green and light-green tesserae suggests different colouring recipes, probably from two different glassmaking centres.

In both turquoise tesserae different layers can be observed, which result from the rough mixture of one transparent and one slightly opaque glass. In the opaque layers, the tin and lead contents are slightly higher (SnO_2 0.4%; PbO 0.4–0.5%) than in the transparent ones (SnO_2 < 0.2%; PbO < 0.3%). The opacity is due to small whitish bubbles and few micro-crystals identified by SEM as euhedral needle-like crystals of tin oxide (SnO_2, cassiterite).

In the presence of oxidised iron, the light-blue colour due to copper tends to shift towards turquoise-green, according to the composition and to the oxidising conditions of the melt. The dark-green tesserae (NAn-C13 and NX-D32) and the light-green tesserae (NAs-C5, NX-D33) are made of a transparent green glass coloured by copper (CuO 1–1.6%) and

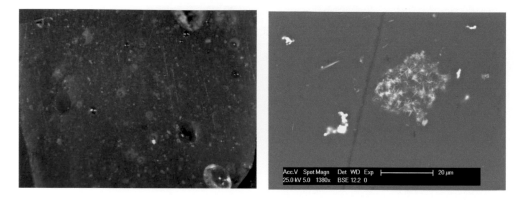

Figure 4.3.19 Polished section of the light-green tessera NAs-C05 – Optical micrograph (left) showing
whitish bubbles and yellow particles of lead stannate; SEM micrograph (right), showing
needle-like SnO2 crystals and rounded particles of lead stannate.

iron (Fe_2O_3 0.4–1.0%). The final colour is the result of the glass colour and the relative
abundance of white and yellow particles (Figure 4.3.10). The SEM investigation has identified two kinds of micro-crystals (Figure 4.3.19). Rounded lead stannate yellow particles are
more abundant in the light-green tesserae, and needle-like cassiterite white crystals similar
in appearance to those observed in the turquoise tesserae are more abundant in dark-green
tesserae. The euhedral aspect of the cassiterite crystals indicate that they result from the
decomposition process of yellow lead stannate particles followed by the dissolution of lead
in the melt and the crystallisation of cassiterite. This is confirmed by the high lead-to-tin
ratio of these tesserae (in the range of 6 to 10) and the positive correlation of these oxides,
generally found in glasses coloured by lead stannate yellow particles.

This pattern is typical of the addition of lead stannate particles to a soda-lime-silica glass
melt and the subsequent thermal decomposition and dissolution of the yellow pigment at
high temperatures to form secondary needle-like cassiterite crystals (Tite *et al.*, 2008).

Yellow (NAn-C11 and NX-D35) and the yellow-green (NX-D34) tesserae are made of a
transparent colourless glass (coloured in green by 0.2% of copper in the NX-D34 tessera)
to which large amounts of yellow pigment were added. The heterogeneous appearance of
these tesserae observed under the optical microscope and SEM (Figure 4.3.20) derives from
the streaks of high-lead glass where yellow particles are abundant. This feature suggests
that these tesserae were made by adding a yellow high-lead intermediate product to the
soda-lime-silica glass melt. Heterogeneity is a common feature of glass coloured with yellow pigments that dissolve easily in the glass melt and decompose at high temperatures. To
avoid losing the yellow colour, the glass has to be worked and cooled as soon as possible
after adding the yellow pigment.

The composition of the pigment particles is similar regardless of the colour of the tesserae.
They are made of lead (PbO 50–64%), tin (SnO_2 24–32%) and silica (SiO_2 10–13%). The
lead-to-tin ratio in the particles is in the range 1.9–3.2. An average composition of the lead
stannate yellow pigments was estimated: PbO 59%; SnO_2 30%; SiO_2 11%. These results
strongly suggest that the yellow particles are artificial pigments prepared following a process
probably quite similar to the preparation of the yellow pigment described in Renaissance

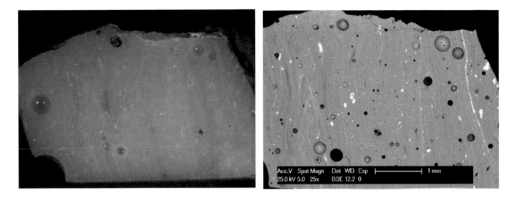

Figure 4.3.20 Micrographs of the polished cross section of a fragment of NAn-C11 yellow tessera observed under the OM (left) and SEM (right).

Venetian glassmaking treatises (Moretti and Hreglich, 1984). In these manuscripts lead stannate pigments were prepared by firing a batch made of lead tin calx (a mixture of lead oxide and tin oxide generally in a ratio 1 to 1), lead oxide and silica. The product was mixed with lead oxide, silica and glass cullet and fired again. Once the yellow pigment was added, the molten soda-lime glass was roughly stirred and rapidly worked. The instability of the pigment in the melt is emphasised in the recipes by recommendations to avoid its dissolution or decomposition and loss of the yellow colour.

Tesserae in a similar range of colours from turquoise to yellow have been used extensively in early Islamic mosaics at Qusayr'Amra (Jordan), and in the mosaics of the Great Mosque of Damascus and the Dome of the Rock in Jerusalem. Instead these colours are not common in Byzantine mosaics such as in Daphni, Hosios Loukas, Monreale and Torcello.

The results of the analyses suggest that the tesserae of this group were manufactured before the end of the 12th century and made with a technology specific to the Levantine area, probably of Islamic tradition.

4.3.3.1.5   RED AND BLACK TESSERAE

Red opaque tesserae in a range of hues from bright to dark brown-red (Figure 4.3.21) and black tesserae with a smooth and bright surface are present in a significant amount in the mosaics of the Bethlehem Basilica.

Two bright-red (NAs-A6 and TRs-A18) and one brown-red (NAs-A7) tesserae were analysed. Their glass is of C1 type coloured with copper (CuO 1.4–1.7%) and iron (Fe_2O_3 2.7–2.9%). Traces of zinc (ZnO 0.1–0.25%) and lead (PbO 0.4–0.8%) suggest the use of a metallurgical by-product (a Cu-Pb-Zn alloy) as a source of copper. The colour and opacity are generated by microspheres of metallic copper. Iron, probably in the form of FeO, was added to the melt as a reducing agent and to promote the formation of the metallic copper microparticles (Freestone *et al.*, 2003).

The differences in colour (bright red and brown red) cannot be explained in terms of chemical composition, for the composition of the three tesserae is quite similar. The brown-red tesserae show dark streaks of transparent green glass, which are absent in the bright-red

Figure 4.3.21 Red bright and red brown tesserae; nave north wall.

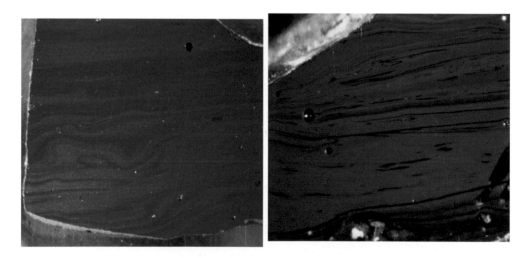

Figure 4.3.22 Optical micrographs of the polished sections of NAs-A6 bright-red (left) and NAs-A7 dark-red (right) tesserae (long side of the micrographs: 2 mm).

ones (Figure 4.3.22). The transparent streaks show the same chemical composition as the opaque red areas, being made of oxidised glass where the copper particles are dissolved.

It is likely that the glassmakers who made the red tesserae for the Nativity Church took advantage of this phenomenon, for instance by mixing the oxidised transparent glass from the surface of the melt with the red opaque glass at the bottom so as to obtain a range of red colour hues.

The black tesserae BEs-C9 and NX-D28 are similar in composition, made of a translucent C1-type glass intensely coloured in amber green by addition of iron (Fe_2O_3 3–3.7%). Dark-brown streaks rich in iron observed in the polished sections confirm the addition of iron. The colour was obtained by maintaining reducing (low oxygen) conditions in the melt so as to favour the formation of the intensely amber-coloured iron-sulphur complex.

4.3.3.1.6 METAL LEAF TESSERAE

The observation of the mosaic areas of the Nativity Church decorated with gold tesserae reveals the presence of a wide range of hues, from cold-light yellow to warm-dark yellow (Figure 4.3.23).

Gold tesserae representing the whole range of hues (taken also from the loose tesserae from the narthex) and one silver tessera were sampled for analysis. These tesserae are made of a thin beaten gold (or silver) leaf (thickness: 0.2–0.6 micrometers for gold, around 1 micrometer for silver) hot fixed between a thin homogeneous glass (*cartellina*, blown glass 0.2–0.5 mm thick) and a thick heterogeneous layer (6–9 mm thick) made of poured molten glass (support). A rough, opaque layer of reddish or grey powder is firmly adhering to the bottom of the support. These characteristics are in agreement with a manufacturing process suggested for previously analysed Byzantine gold-leaf tesserae (Verità and Rapisarda, 2008). Thin blown-glass fragments were first cut (the *cartellina*). A beaten metal leaf was laid upon and molten glass was then poured on the leaf. The obtained product was then turned rapidly upside down by placing the still-plastic layer of poured glass on a flat surface covered with an anti-adhesion powder. A flat tool was used to press the sandwich in order to ensure adhesion between the layers and obtain slabs from which the tesserae were cut.

The aspect of these tesserae is the result of the colours of the metal leaf, the *cartellina* and the support glass. In fact, the thin metal leaves are discontinuous and semi-transparent because of tears caused by pressing during the forming of the slab, so that the aspect of the tessera is influenced also by the colour of the support glass.

Figure 4.3.23 Gold-leaf tesserae, nave, northern wall, next to an angel.

Glass and metal leaf compositions of the examined tesserae are reported in Table 4.3.3. In some tesserae the *cartellina* resulted to be similar in composition to the support (tesserae indicated with (sc)). In others the two layers are different and the composition of the *cartellina* (c) and support (s) are reported separately. Five tesserae (two from the nave and three from the narthex) are made with an N2-type glass; four tesserae from the narthex are made of C1-type glass, and two tesserae (one from the bema and one from the transept) are made of C2-type glass.

In the cold-light yellow gold tesserae both the support and the *cartellina* are slightly coloured in yellow (indicated as *colourless* in Table 4.3.3) as the result of the iron content (Fe_2O_3 0.4–0.9%) and the decolouration by addition of manganese (MnO 0.7–1.7%). Decolouration is based on the oxidation of iron into a faint yellow Fe^{3+} by addition of manganese (chemical decolouration). When a slight excess of manganese is added, it colours glass in purple neutralising the residual yellow colour of Fe^{3+} (physical decolouration).

In the warm yellow brown gold tesserae the glass of the *cartellina* (or of the support) is intensely coloured. This pattern is observed in the natron type as well as in soda plant ash–type gold tesserae. In one tessera from the narthex NX-OE27, both the *cartellina* (brown) and the support (olive green) are intensely coloured.

The brown colour is due to iron (Fe_2O_3 0.4–0.8%) and to significant amounts of manganese (MnO 1.5–3.2%). The manganese–iron-brown colour in glass depends on the amount of these elements and on their oxidation state. In some tesserae the colourless glass has iron (Fe_2O_3 0.4–1.1%) and manganese (MnO 0.7–3.3%) contents comparable or even higher than the brown glass. The manganese–iron combination is highly sensitive to the oxygen level in the glass melt, and the final glass colour is difficult to stabilize with respect to furnace atmosphere fluctuations (Volf, 1984). Therefore, it is not a surprise to have in one tessera different layers, some oxidised ones where the glass is brown and others less oxidised where glass is colourless (Figure 4.3.24). Clearly, the Byzantine glassmakers succeeded in governing this delicate redox equilibrium so that they managed to obtain different colour intensities

Figure 4.3.24 Optical micrograph of the polished cross section of a gold tessera with layered pattern of the brown/colourless support.

from a same melt and, consequently, a wide colour range for gold-leaf tesserae. A similar colouring technique was identified in tesserae from the archaeological site of Qusayr'Amra (Jordan) dated to the first half of the eighth century (Verità and Santopadre, 2017) and then in other Levantine mosaics (Neri *et al.*, 2016), a feature never observed in gold tesserae from occidental mosaics.

The leaf of the gold tesserae is made of gold–silver alloys. A similar silver content (Ag 2% to 12%, concentration expressed in wt% of the element) was found in the natron type as well as in the soda plant ash–type tesserae. The use of different gold–silver alloys may have been an expedient to widen the colour range of the tesserae. Nevertheless, the relationship between colour and composition demonstrates that the colour of gold alloys containing Cu < 1% and Ag < 4% is in the same red-yellow range as pure gold (Au 100%; Rapson, 1990). The colour turns to yellow with increasing silver contents (4% < Ag < 12%) and to green-yellow in gold alloys containing silver in the range 12% < Ag < 28% (and Cu < 1%). Therefore, if the gold–silver alloys were used as a means to obtain a large variety of gold hues, it is not evident why the manufacturers of the leaves did not use alloys with a silver higher than 12% to widen their palette.

More likely, the composition of the gold alloy was related to the availability of gold to prepare the leaves. In this context, circulating gold coins were a handy gold source. A gold leaf made by beating circulating coins was demonstrated for a number of tesserae from first- to ninth-century mosaics in Rome (Neri and Verità, 2013). Similarly, by analysing gold-leaf tesserae from Levantine mosaics dated to between the fourth and 12th centuries, the authors found that their composition is quite variable from the eighth century onwards with a silver content up to Ag 10% (Neri *et al.*, 2016). The same variability was revealed by the analyses of the circulating gold coins from the Levantine area of the same period (Morrisson, 2002).

In general, silver-leaf tesserae are relatively rare in mosaics. These are delicate, structurally unstable tesserae and, therefore easily weathered. Silver is a less ductile metal than gold, so that silver leaves are thicker and, consequently, the different expansion coefficients of glass and silver give rise to increased stress. Moreover, silver oxidises relatively easily to form black silver sulphide in the presence of humidity and polluted environments. This phenomenon starts at the edge of the tesserae and propagates concentrically to the centre, thus causing the loss of brightness, the detachment of the *cartellina* and at the end the loss of the silver leaf. In the mosaics of the Nativity Church the silver-leaf tesserae are unusually abundant and well preserved, considering the long time they have been in situ. The characteristic forms of weathering are visible only in few tesserae (Figure 4.3.25). This is probably due to the fact that the indoor environment has remained poorly aggressive for glass mosaic tesserae.

One silver tessera was sampled for the analyses (NAn-A8); both the *cartellina* and the support are made of colourless glass. Anyway, the examination of the mosaics revealed the use of coloured glass for the support and/or the *cartellina* of the silver tesserae to extend the chromatic range, exactly as discussed above for the gold tesserae. The glass of the NAn-A8 tessera is of natron N2 type; a slightly different manganese content was observed for the *cartellina* (MnO 1.4%) and support (MnO 1.7%). The metal leaf is composed almost exclusively of silver with traces of copper (Cu 0.7%).

In conclusion, to prepare metal-leaf tesserae in a wide range of colours, the support and/ or the *cartellina* glass were made with deliberately coloured glass, a technique specific to Levantine mosaics. In fact, a voluntary colouring of glass has been ascertained in Europe only for a few silver-leaf tesserae from the fifth century mosaics of Santa Sabina, Rome (Verità, 2017).

Figure 4.3.25 Silver-leaf tesserae (nave northern wall), some showing a brown-coloured support. On the left, some weathered tesserae (concentric corrosion).

4.3.3.1.7   REUSED TESSERAE

The common practice of reusing glass tesserae from dismantled ancient mosaics was a way of cutting costs and also of providing those colours and types of tesserae that could not be obtained at the time. Indeed, it must be considered that the palette of tesserae used in ancient mosaics was in part determined by the ability of the glassmakers to obtain certain colours and that the availability or not of certain colourants and opacifiers influenced the final appearance of a mosaic.

The technological characteristics of the glass tesserae recovered from the materials accumulated on the vaults of the narthex (N1, N2 and C1 glass, use of lead stannate pigments, etc.) are similar to those of the tesserae of the mosaics of the basilica, except for a few tesserae made with N2 natron-type glass (Egypt).

Are the narthex loose tesserae from early mosaics in the church (Figure 4.3.26)? Or from mosaics in other sites? Or left over tesserae on completion of the mosaics of the basilica? Until now the analytical data cannot answer these questions for now.

4.3.3.2   Summary

The analyses of glass tesserae from the wall mosaics of the Nativity Church of Bethlehem disclose a complex situation that can be attributed to the variety of materials and manufacturing techniques used. This situation is not surprising, for the presence of glass tesserae different in composition, origin and production period is a rule rather than an exception in ancient mosaic decoration. On the other hand, mosaics of the basilica were made in a period of great changes in glassmaking that began in the eighth to ninth centuries and came to an end only in the 13th century. The analytical data can be summarised as follows:

Figure 4.3.26 Loose tesserae from the deposits on the vaults of the narthex.

The tesserae were partly made with a natron-type glass (N1 and N2 groups), and partly with a soda plant ash glass (C1 and C2 groups). Gold-leaf tesserae were made with both N2 glass and C1-C2 glass. N1 glass tesserae are only present among the loose tesserae of the narthex. Regardless of the type of base glass, the technique is the same: glass coloured in brown and gold-leaf composition (gold–silver alloy) varying in the same range (Ag 2–12%). This could mean that gold and silver tesserae were made in the same Levantine workshop that took supplies of raw glass made following two different recipes: natron-type glass and soda plant ash–type glass.

Blue, red and black tesserae are exclusively made of C1 soda plant ash glass. Probably, also other non-analysed tesserae coloured in brown, purple, green and grey and opacified with ground quartz fall into this group. Conversely, tin opacified turquoise to yellow tesserae were made with N1 and N2 natron-type glass.

These data suggest that the Bethlehem Nativity Church glass tesserae may come from two-three workshops. One workshop specialised in the production of quartz opacified coloured tesserae, soda plant ash–type glass. Another workshop specialised in the synthesis of the lead stannate pigment and in making turquoise to green-yellow tesserae with natron glass. The manufacture of metal-leaf tesserae (made both with natron type and soda plant ash–type glass) requiring specific knowledge and skill, it could have taken place in a special-ised site or in one of the workshops where the coloured tesserae were made. The analytical data suggest a location of the workshops in the Levantine (Byzantine and Islamic) area. The absence of tesserae manufactured in the West has been ascertained at the same time. The reuse of ancient tesserae opacified with bone ash recovered from dismantled mosaics dated between the fifth and the 10th centuries in the Levantine area (maybe even from the basilica) has been also attested. As far as we know today, the coloured tesserae made with C1-type glass, intermediate between natron glass and soda plant ash glass were made and used in a restricted period between the 11th and the 12th centuries.

The mosaicists who worked in all the decorated areas of Nativity Church walls (nave, transept, bema) had the same palette of glass tesserae available, made following the above-mentioned recipes. This suggests that all the mosaics were assembled in the same period of time.

Not all the colours in the mosaics of the Nativity Church were obtained with glass tesserae. The making of glass tesserae is a difficult process for a variety of reasons. Some colours were easy enough to produce but required rare and/or very expensive raw materials; others were far more complex, requiring experience and skill. These difficulties would encourage the use of other materials. The use of stone tesserae for the white and flesh-tone tesserae in the Nativity Church mosaics is characteristic of Byzantine mosaic decoration in general. Flesh-tone glass tesserae were (and are) produced by adding gold ruby glass to a white opaque glass. This sophisticated technology involving gold nanoparticles was known to the Roman glassmakers of the early centuries of the Christian era (Verità and Santopadre, 2010). Later this recipe was lost, and for this reason stone tesserae were used for flesh tones in Byzantine mosaics and in the mosaics of the Nativity Church.

On the other hand, quartz and bone ash were inexpensive opacifiers, suitable for lightening (and obtaining three or four hues) coloured glass tesserae, but they are much less effective than the very expensive tin and antimony to make a white opaque glass. Therefore, it is probably for economic reasons that the white tesserae of Byzantine mosaics and of the Nativity Church are made of stone and not of glass.

No restoration or repair using tesserae made in later times has been identified. This confirms that the mosaics have undergone no refurbishment after their assembly or, if any, ancient tesserae were reused. An additional explanation is the fact that the indoor environment in the basilica has remained (and still seems to be) not particularly aggressive over time as regards the mosaic glass tesserae. This is confirmed by the relatively good state of preservation of the delicate silver-leaf tesserae.

References

Antonelli, F. & Lazzarini, L. (2015) An updated petrographic and isotopic reference database for white marbles used in antiquity. *Rendiconti Lincei Scienze Fisiche e Naturali*, 26(4), 399–413.

Arletti, R., Fiori, C. & Vandini, M. (2010) A study of glass tesserae from mosaics in the monasteries of Daphni and Hosios Loukas (Greece). *Archaeometry*, 52(5), 796–815.

Bacci, M. (2015) Old restorations and new discoveries in the nativity church, Bethlehem. *Convivium*, 2(2), 37–59.

Bacci, M. (2017) *The Mystic Cave: A History of the Nativity Church in Bethlehem*, Viella, Masaryk University Press, Rome, Brno.

Bacci, M., Bianchi, G., Campana, S. & Fichera, G. (2012) Historical and archaeological analysis of the church of the nativity. *Journal of Cultural Heritage*, 13, e5–e26.

Brill, R.H. (2008) Chemical analyses of some glass mosaic tesserae from Hosios Loukas and San Nicolo di Lido. *Musiva and Sectilia*, 5, 169–190.

Folda, J. (1995) *The Art of the Crusaders in the Holy Land 1098–1187*, Cambridge. p. 361.

Folk, R.L. (1959) Practical petrographic classification of limestone. *American Association of Petroleum Geologists Bulletin*, 43(1), 1–38.

Freestone, I.C. (2005) The provenance of ancient glass through compositional analysis. *Materials Research Society Symposia Proceedings*, 852, 1–14.

Freestone, I.C., Bimson, M. & Buckton, D. (1988) Compositional categories of Byzantine glass tesserae. *Annales AIHV*, 11, 271–279.

Freestone, I.C., Stapleton, C.P. & Rigby, V. (2003) The production of red glass and enamel in the Later Iron Age, Roman and Byzantine periods. In: Entwistle, C (ed.) *Through a Glass Brightly: Studies in Byzantine and Medieval Art and Archaeology Presented to David Buckton*. Oxbow. pp. 142–154.

Gratuze, B. & Barrandon, J.N. (1990) Islamic glass weights and stamps: Analysis using nuclear techniques. *Archaeometry*, 32, 155–162.

Henderson, J. & Roe, M. (2006) Technologies in transition: Torcello glass tesserae, primary glass production and glass trade in the medieval Mediterranean. *Arte Medievale*, 2, 120–140.

Henderson, J., McLoughlin, S.D. & McPhail, D.S. (2004) Radical changes in Islamic glass technology: Evidence for conservatism and experimentation with new glass recipes from early and middle Islamic Raqqa, Syria. *Latin Archaeometry*, 46, 439–468.

James, L. (2017) *Mosaics in the Medieval World: From Late Antiquity to the Fifteenth Century*. Cambridge University Press, Cambridge.

Kühnel, G. (1988) *Wall Painting in the Kingdom of Jerusalem*. Berlin.

Lazzarini, L. (2015) Il reimpiego del marmo proconnesio a Venezia. In: Centanni, M. & Sperti, L. (eds.) *Pietre di Venezia, spolia in se, spolia in re*. Ariccia, Rome. pp. 135–157.

Marii, F. & Rehren, T. (2009) Archaeological coloured glass cakes and tesserae from Petra church. In: Janssens, K., Degryse, P., Cosyns, P., Caen, J. & Van't Dack, L. (eds.) *Proceedings of the 17th AIHV Conference*. University Press Antwerp, 295–300.

Moretti, C. & Hreglich, S. (1984) Opacification and colouring of glass by the use of Anime. *Glass Technology*, 25(6), 277–282.

Morrisson, C. (2002) Byzantine Money: Its production and circulation. In: Laiou, A.E. (ed.) *The Economic History of Byzantium: From the Seventh through the Fifteenth Century*. Dumbarton Oaks, Washington. pp. 909–966.

Neri, E. & Verità, M. (2013) Glass and metal analyses of gold leaf tesserae from 1st to 9th century mosaics: A contribution to technological and chronological knowledge. *Journal of Archaeological Science*, 40, 4596–4606.

Neri, E., Verità, M., Biron, I. & Guerra, M.F. (2016) Glass and gold: Analyses of fourthe12th centuries Levantine mosaic tesserae: A contribution to technological and chronological knowledge. *Journal of Archaeological Science*, 70, 158–171.

Neri, E., Verità, M. & Biron, I. (2018) New insights of Byzantine glass technology from lost mosaics of Hierapolis (Turkey): PIXE/PIGE and EPMA analysis of glass tesserae. *Journal of Archaeological and Anthropological Sciences* (In press).

Phelps, M., Freestone, I., Gorin-Rosen, Y. & Gratuze, B. (2016) Natron glass production and supply in the late antique and early medieval Near East: The effect of the Byzantine-Islamic transition. *Journal of Archaeological Science*, 75, 57–71.

Rapson, S.W. (1990) The metallurgy of the coloured carat gold alloys. *Gold Bulletin*, 23(4), 125–133.

Santopadre, P., Bianchetti, P. & Sidoti, G. (2017) Le malte di mosaici medievali romani. Composizione e struttura. In: Andaloro, M. & D'Angelo, C. (eds.) *Mosaici medievali a Roma attraverso il restauro dell'ICR 1991–2004*. Gangemi, Rome.

Silvestri, A., Tonietto, S., Molin, G.M. & Guerriero, P. (2012) The palaeo-Christian glass mosaic of St. Prosdocimus (Padova, Italy): Archaeometric characterisation of tesserae with antimony- or phosphorus-based opacifiers. *Journal Archaeological Science*, 39, 2171–2190.

Stern, H. (1936) Les représentations des Conciles dans l'église de la Nativité à Bethléem. *Byzantium*, 11, 101–152.

Tite, M., Pradell, T. & Shortland, A. (2008) Discovery, production and use of tin-based opacifiers in glasses, enamels and glazes from the late iron age onwards: A reassessment. *Archaeometry*, 50, 67–84.

Uboldi, M. & Verità, M. (2003) Scientific analyses of glasses from Late Antique and Early Medieval archaeological sites in northern Italy. *Journal of Glass Studies*, 45, 115–137.

Verità, M. (2010) Glass mosaic tesserae of the Neonian Baptistry in Ravenna: Nature, origin, weathering causes and processes. In: Fiori, C. & Vandini, M. (eds.) *Proceedings of the Conference: Ravenna Musiva*. Ante Quem, Ravenna. pp. 89–103.

Verità, M. (2017) Studio delle tessere vitree di mosaici medievali romani. Tecnologia e degrado. In: Andaloro, M. & D'Angelo, C. (eds.) *Mosaici medievali a Roma attraverso il restauro. Storie di otto cantieri*. Gangemi, Rome.

Verità, M. & Rapisarda, S. (2008) Studio analitico di materiali musivi vitrei del XII–XIII secolo dalla Basilica di Monreale a Palermo. *Rivista della Stazione Sperimentale del Vetro*, 38(2), 15–29.

Verità, M. & Santopadre, P. (2010) Analysis of gold-colored ruby glass tesserae in Roman church mosaics of the fourth to 12th centuries. *Journal of Glass Studies*, 52, 1–14.

Verità, M. & Santopadre, P. (2017) Scientific investigation of glass mosaic tesserae from the 8th century AD archaeological site of Qusayr' Amra (Jordan). *Bollettino ICR, Nuova Serie*, 32, 5–20.

Verità, M. & Zecchin, S. (2012) Scientific investigation of the Byzantine glass tesserae from the mosaics of the south Chapel of Torcello's Basilica, Venice. In: Despina Ignatiadou, D. & Antonaras, A. (eds.) *Proceedings of the 18th AIHV Conference*. Thessaloniki. pp. 315–320.

Volf, M.B. (1984) Chemical approach to glass. *Glass Science and Technology*, 7, 340–346, Elsevier.

4.4 The photogrammetric survey of the mosaics

G. Caratelli and C. Giorgi (CNR)

The current landscape of survey techniques in the field of cultural heritage is so broad, varied and always evolving that it requires professionals working with graphic documentation in archaeology, restoration or cataloguing projects to be constantly informed and updated. However, when working site schedules and deadlines need to be met, and considering clients' requirements for reliable results, in both the private and public sectors, anyone involved in surveying has to select the best suitable approaches, by taking into account the characteristics of the cultural object, the context, the equipment and financial resources, as well as the aims to be achieved.

The wall mosaics in the central nave and transepts in the Church of the Nativity in Bethlehem, before the careful restoration works, were dark and almost illegible and, in some areas, even covered by more recent layers. The multidisciplinary project team in charge of cleaning the mosaics required a very detailed survey, accurately and realistically reproducing the entire decoration to be restored. This essential graphic representation was used to plan restoration works and define the framework for individual operations. It was also useful for developing thematic maps on the state of conservation and deterioration, for orienting and supporting the stylistic and historical analysis, as well as studies on how the mosaic was created. Furthermore, it had to include accurate information on quantity, distribution and properties of the materials used.

The main objective of the survey, therefore, was a graphic elaboration of the entire mosaic decoration, so that its actual extension and punctual distribution could be appreciated without overlooking the tiles' shape, quantity and materials. In addition, it was necessary to complete the data acquisition phase as rapidly as possible in order to avoid inconvenient overlapping with the restoration works already underway and, especially, to provide this necessary survey as a support and analysis tool (Ippoliti, 2000) for the work carried out by restorers.

Had the authors been appointed this task about 20 years ago, their range of options would have been much more restricted, and they would probably have adopted an entirely manual survey technique. This would have consisted in making a real-scale copy of decorated flat surfaces (Medri, 2003); in the case of the Nativity Church in Bethlehem, the mosaic would have been traced by marking each tile profile on transparent plastic sheets, previously laid on the mosaic itself. After completing the copy, through an appropriate rescaling of each sheet, and their reassembly, the entire representation of the mosaic decoration could have been obtained. Clearly, this type of survey, in addition to causing a number of technical difficulties and metric inaccuracies, would have taken a very long time to be performed in both the acquisition and processing phases.

For these reasons, considering also the need for a detailed representation of the vast mosaic surface (more than 120 square metres) without simplifying or sampling, in a short period of time and, especially, in the particular conditions described afterwards, we decided to use digital photogrammetry as a basis for tessera-by-tessera vector drawings in a CAD (Computer Aided Drafting) environment (Birocco, 1996; Bianchini, 2008). It was, therefore, an ambitious objective, since the mosaic decorations to be restored – located in the central nave (north wall, sectors A–H, Tables 4.3.1, 4.3.2; south wall, sectors A–D, Table 4.3.3) and in the north and south transepts (Tables 4.4.4, 4.3.5) – included a number of tesserae ranging from 13000 to 15000 per square metre of decoration.

For this reason, the acquisition phase was designed to obtain the best possible results with the fewest human and economic resources, the latter being mostly reserved for the graphic restitution phase. In fact, the most onerous operation consisted in elaborating vector drawings for identifying each tessera. On the one hand, a photogrammetric basis correctly representing reality (a "digital clone") was needed; on the other hand, a drawing interpreting and explaining it was essential. This would be achieved by using specifically selected and classified sections so as to obtain as much information as possible by transforming the photogrammetric object into multiple classified objects, i.e. the mosaic tesserae. We then chose a specific survey technique that enabled us to support and speed up the creation process of these drawings: two-dimensional photogrammetry based on image rectification. This technique, compared with other survey methods (such as laser scanning survey), can both provide very high metric reliability levels and is convenient in terms of implementation and management.

Photogrammetric theory is based on the principles of projective geometry and, specifically, on the solution for the "inverse problem of perspective" (Paris, 2014). Every photograph, in fact, can essentially be considered as a perspective (Paris, 2014) once distortions of the projected image caused by light travelling through the camera lens are taken into account. Therefore, it is possible to retrace (through analogue or analytic means) the shape and size of the photographed object. If the object is flat, one photograph is sufficient (monoscopic photogrammetry), while at least two photos are needed if one wants to recover the object's depth as well (stereoscopic photogrammetry). During the 1990s, the introduction of digital photography caused a veritable revolution in this field, rapidly transforming photogrammetry in one of the most valid, common and used survey techniques (Carpiceci, 2012). The latest photo-modelling applications, using computer vision image-based algorithms, represent a further development in photogrammetry and in 3D modelling, which has rendered the consolidated laser scanning survey less attractive (mostly for cost-related reasons). Since the introduction of digital photography, the technique based on image rectification has become increasingly common, also in the cultural heritage context, taking advantage of open-source software development.

In the case of the Bethlehem mosaics, the free software RDF, developed by the Photogrammetry Laboratory of the Iuav University of Venice, was used. This software allows us to rectify an image of a flat object if the Cartesian coordinates (real-world coordinates) of at least four points located on the object's plane are known. This processing phase is quite rapid and automatic; it is necessary to recognise and collimate such four known points (called homologous points) in the image in order to deduce the pixel coordinates to be assigned to the real ones. In this way, the software is able to generate a new image by calculating real-world coordinates of all other points, provided that they are placed on the same plane.

In brief, the five phases of the survey project were as follows:

1   arrangement and measurement of control points on the mosaic for image rectification process;
2   acquisition of digital photograms in order to generate high-resolution images;
3   image rectification process;
4   photogrammetric representation of the entire mosaic, divided in sectors;
5   elaboration of vector drawings in CAD environment.

Phase I

Firstly, we selected locations for control points, targets distributed homogeneously on the mosaic surface and visible on the photograms. Targeting operations are an essential part of photogrammetric mapping, which has to be considered prior to establishing a control survey. We planned target placement in order to divide every decorated sector in rectangular-shaped areas, where the targets were the vertexes. The arrangement of this dense and regular point network was projected to create a survey basis, covering the entire surface, and to design the image acquisition phase (Figure 4.4.1).

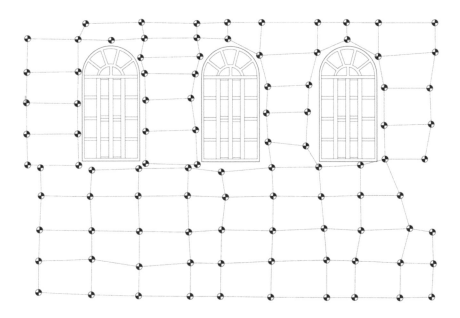

Figure 4.4.1 Central nave, north wall particular. Distribution of targets (in red colour) to divide the mosaic in areas and create a dense point network.

Every rectangular-shaped area corresponded to a photogram, in which at least four targets had to be visible. The distance between each target, generally less than 1 metre, was closely related to that between the mosaic surface and the camera and, therefore, to the desired image resolution, which will be explained in detail in Phase 2.

Targets were uniquely identified by number and colour, placed on the surface with adhesive paper. They were surveyed using a measuring tape, by making trilaterations for determining the local coordinates of every point. The measures were transferred in a CAD environment and constantly controlled in order to limit relevant error propagation in the range of millimetres. Unfortunately, using topographic instruments for measurements, such as Total Station, was not possible because of the worksite conditions. There were many irremovable obstacles and, especially, the worksite scaffoldings were constructed to support a suspension floor on which the professionals could work. This suspension floor, built between the floor and the ceiling of the church, enabled working close to the mosaic, given the latter's location on the walls, but at the same time did not provide a stable enough support to enable levelling up the topographic instrument.

Phase 2

The photograms were acquired through a digital reflex camera, equipped with a 10.2-megapixel CCD sensor (size mm. 23.6 × 15.8; pixel 3872 × 2592) and with a 35 mm fixed lens. Obviously, this choice, especially regarding the lens, was not arbitrary but was determined by the high-definition image we wanted to achieve. Moreover, the small size of tesserae, with a side length of maximum 1 cm, required a shooting distance between 1.2 m and 1.5 m, which resulted in an image resolution of 4 or 5 pixels/mm.

Each photogram was shot so that the optical axis was as much as possible orthogonal to the mosaic surface. Where scaffolding tubes blocked the view, it was necessary to acquire oblique images in order to survey the rear area. Lastly, considering the low natural illumination inside the church, the camera flash was systematically used to reduce exposition times and increase image sharpness and clarity.

Phase 3

Before image rectification processing, all photograms were processed using the Adobe Camera Raw software, an Adobe Photoshop plug-in, in order to remove inevitable distortions, typically occurring in images captured with a camera. Indeed, it is known that a series of optical aberrations are introduced when light passes through the lens (most commonly the pincushion and barrel distortion). These aberrations can substantially affect the correct geometrical projection of the object reproduced on the sensor surface.

Before the introduction of digital photography, this issue was resolved by using metric cameras, but now the aberrations – even the most significant ones affecting lenses with smaller focal lengths – can be corrected by specific software processing the original digital image, by taking account of the typical distortion parameters for every lens type. The result is a new digital image that can be finally used for metric purposes.

The image rectification processing phase began after completing the essential preliminary operation described above. The aim of this phase was transforming every undistorted image in a central perspective view of the captured mosaic area. In a central perspective view, indeed, the projection plane is parallel to the object's flat surface one wants to represent.

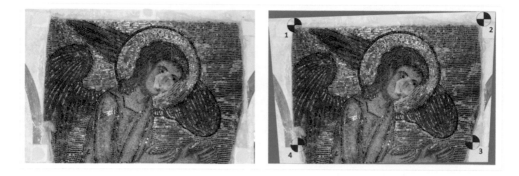

Figure 4.4.2 Central nave, north wall particular. Original photogram with the representation of an angel (left side); image rectification processing (right side) that shows the four control points (red targets) lying on the plane to be rectified and the distortions for optical aberrations (orange areas).

This means that all points lying on that same surface are in the same scale, such as in an orthographic view (orthographic projection). The software can perform this process by using either the geometrical or the analytical method. The former is used only in architecture, when the vertical and horizontal directions on the flat surface can be recognised. The latter, instead, is used in all other cases and, obviously, is the most common in cultural heritage applications, where the objects are irregularly shaped.

In relation to a mosaic flooring, for example, the geometrical method could only be applied to a geometrical decoration arranged in an orthogonal pattern. Instead, every image of the Bethlehem mosaic, decorated almost exclusively with figurative designs, was processed by applying the analytical method, which required the local coordinates of at least four control points lying on the plane to be rectified, properly positioned at the ends of every photogram. The software calculated the transformation projective parameters and resampled all photograms by connecting the coordinates of the control points to the coordinates expressed in pixels of the same points visible in the image, with the aim of producing new metrically correct images (Figure 4.4.2).

Phase 4

Finally, in order to obtain a unique ortho-photo for each surveyed sector, every rectified image was imported into a CAD environment and was positioned and scaled based on the geo-referencing information contained in the file created by the software, such as the end coordinates, size in pixels and metres and size of every pixel and resolution of the rectified image.

Phase 5

After producing an accurate metric and photorealistic basis of the entire mosaic decoration, the final surveying phase began. The processed ortho-photos, keeping their metrical value, were common raster bitmap images composed by a range of square-shaped pixels,

Figure 4.4.3 Central nave, north wall particular. Ortho-photo and vector drawing of every tesserae.

each of which was assigned to a specific colour. Clearly, this type of image provided many data about the mosaic, not limited to the metric point of view. However, this information (for example, the total number of tiles used) was not immediately or automatically available.

Therefore, in order to convert the photogrammetric survey into a flexible instrument of both quantitative and qualitative analysis and understanding, it was necessary to proceed with the vector "translation" of every tessera by exactly positioning it on the surface plane and changing it into an element with its own identity, classifiable on the basis of specific features (Barberini, 2006). Moreover, the transition from raster to vector file gave a further significant advantage. The main property of raster images, indeed, is resolution, measured by the ratio between the total number of pixels and the image size, that is the number of pixels included in a certain unit of measurements: the higher this ratio, the higher the image resolution and, consequently, the image quality. Therefore, if the image size is increased, the image loses resolution and becomes grainy, since every pixel becomes visible. Vector graphics, instead, is based on geometrical shapes – lines, points, curves and polygons – described by mathematical equations, and therefore they can be infinitely enlarged while keeping the same resolution. Indeed, vector drawings have greater representation versatility. They can be visualised or printed in any scale: the maximum enlargement depends only on the detail level achieved during the graphic processing phase, while the raster image printing always depends on its resolution.

Furthermore, every tessera – in addition to being identified by a closed polygon – was classified according to colour and assigned to a specific layer. This procedure enabled a better understandability of the file and also contributed to conveying essential and precise metric and spatial information, including the size of the tesserae and their distribution on the reference plane, as well as data concerning the materials (Figures 4.4.3, 4.4.4).

Figure 4.4.4 Transept, south wall particular. Ortho-photo and vector drawing of every tesserae in one square meter.

Conclusions

Surveying and data acquisition techniques for studying mosaic surfaces have remarkably evolved over a few decades, but the usage and function of the final results remained unchanged. Indeed, while until recently the almost exclusively used method was manual direct survey by tracing, current techniques using laser scanner and digital photogrammetry are widely adopted in order to create realistic and metrically accurate reproductions of any decorated object.

The distinguishing factor for these techniques, in addition to precision levels and acquisition timings, is the type of representation that can be obtained and, therefore, the analysis that can be performed. In restoration works of mosaic surfaces, the application of manual tracing survey for reproducing tesserae in a 1:1 scale is still widely attested, but it is only used in the documentation of small areas due to its difficulties, lack of precision and very long timings (Biscontin & Driussi, 2002; Andaloro & D'Angelo, 2017).

The main purpose of this survey technique is to verify the employment of "contours" (or outlines) as a means to transfer the image on the surface to be decorated, before laying tesserae on the setting bed. In fact, drawing every tessera enables comparison and analysis of their distribution and orientation, which are decisive factors for technical and stylistic studies.

On the other hand, surveys using laser scanner and photogrammetric data are applied to produce 3D representations of masonry structures with decorated surfaces or to create orthographic views for a general idea of the depiction and its composition, where only the outlines of the figures are converted in vector format.

Therefore, even with these new survey methods, the vector representation of all tesserae is usually limited to the more relevant sections. As a consequence, the graphical representations obtained with these techniques – mainly thematic plates in various colours – only approximately show the proportions of the analysis, studies and works performed on the mosaics. For example, one can roughly perceive whether some materials were used more than others, but one cannot quantitatively and precisely analyse all the materials and their distribution. Or one can outline the areas affected by restoration and those where materials were sampled but without being able to accurately define the size of the surface being studied. In the specific case of the Bethlehem mosaics, considering the expedited acquisition of photograms to produce ortho-photo mosaics of each sector, the authors thought it appropriate and necessary to perform a complete vectorisation of all mosaic surfaces in a CAD environment. Currently, although research on new software capable of automatically converting raster images into vector files is underway (Felicetti et al., 2018), manual computer-assisted drawing is the only effective way to extract meaningful information from ortho-photos (Carbonara, 2012). This mainly quantitative information can gain complexity, depending on specific needs, through the addition of qualitative metadata to merely geometric data. The information becomes available and can be therefore analysed, compared, schematised or simply conveyed, only after this laborious extraction. Conversely, if surveying operations were exclusively reduced to an exact reproduction of reality, by means of extremely accurate ortho-photos or fascinating three-dimensional models, we should admit that graphical restitution, and therefore drawing, had become obsolete as a privileged means of analysis and knowledge.

References

Andaloro, A. & D'Angelo, C. (2017) *Mosaici medievali a Roma attraverso il restauro dell'ICR: 1991–2004*, Gangemi: Roma.

Barberini, C. (2006) *AutoCAD e il rilievo archeologico digitale*. Morlacchi Editore: Perugia.

Bianchini, M. (2008)*Manuale di rilievo e di documentazione digitale in archeologia*. All'Insegna del Giglio: Roma.

Birocco, C.A. (1996) Raddrizzamento digitale applicato al rilievo dei beni culturali, *Archeologia e Calcolatori*, 7, 73–77.

Biscontin, G. & Driussi, G. (2002)*I mosaici: cultura, tecnologia, conservazione: atti del Convegno di studi, Bressanone, 2–5 luglio 2002*. Arcadia ricerche: Marghera, Venezia.

Carbonara, G. (2012) Disegno e documentazione per il restauro: un impegno interdisciplinare. *Disegnare con*, 5, 10. University of L'Aquila.

Carpiceci, M. (2012) *Fotografia digitale e Architettura. Storia, strumenti ed elaborazioni con le odierne attrezzature fotografiche e informatiche*. Aracne: Roma.

Felicetti, A., Albiero, A., Gabrielli, R., Pierdicca, R., Paolanti, M., Zingaretti, P.& Malinverni, E.S. (2018) Automatic Mosaic Digitalization: A Deep Learning approach to tessera segmentation. *2018 IEEE International Conference on Metrology for Archaeology and Cultural Heritage Proceedings, Cassino, University Campus, 22–24 ottobre*, IEEE.

Ippoliti, E. (2000) *Rilevare. Comprendere, misurare, rappresentare*. IEEE: Roma.

Medri, M. (2003) *Manuale di rilievo archeologico*. GLF editori Laterza: Roma, Bari.

Paris, L. (2014) *Dal problema inverso della prospettiva al raddrizzamento fotografico*. Aracne: Roma.

4.5   State of conservation

S. Sarmati, N. Santopuoli and E. Concina

The conservation history of the basilica is very complex, and the series of changes, due in part to the historical context of the territory, have had strong repercussions on the appearance and state of conservation of the mosaics.

Historical evidence reveals that the mosaics had already begun to show signs of damage as early as the 14th century, as a large number of tesserae had come loose and fallen off.

In 1406, a pilgrim visiting the Nativity Church made note of the deterioration: various portions of the mosaic in the apse had already been lost, whereas others, such as the procession of Christ's ancestors, were dark and hard to see.[15]

Additional observations of the decay of the mosaic facing were made over the centuries.

In 1699, Giovanni Giustino Ciampini made an engraving of the entire north wall of the nave. The image is not a perfect representation of the decoration; all of the angels are depicted in the same way, and none of them are drawn in the right position.

Portions of the mosaic had probably already been lost, so Ciampini recreated a hypothetical, albeit rather accurate, vision of the decoration as a whole based on fragments which presumably would have been much larger than those we see today.

In 1842, the basilica, which had already been greatly damaged by large infiltrations of rainwater from the roof, underwent a complete restoration. The basilica that we see today is the product of this renovation work, which completely renewed the church's appearance: the walls were completely covered with white plaster, the last erratic tesserae on the walls were eliminated, and the more extensive mosaic fragments were framed, as if each image were independent from the others.

In 1983, in order to do a systematic study of the decorations in the basilica, the scholar Gustav Kühnel, assisted by a Greek art restorer, cleaned the surface of the mosaic fragments by removing the black smoke while leaving the touch-ups and plaster work done during the previous restoration.

Figure 4.5.1 Nave. Wall mosaic. Cracks.

Much of the degradation can be attributed to the infiltration of rainwater, which, unfortunately, continued until a short time ago.

The analysis of the surface carried out during the planning phase indicated extensive adhesion defects in the preparatory layers as well as some deformations and long cracks on the north wall of the nave and in the transept (Figure 4.5.1). The layers of preparatory mortar were not very cohesive and did not adhere well to each other nor to the wall structure. The gaps on the surface were caused by tesserae that had fallen off, revealing the coloured backgrounds of the setting bed (Figure 4.5.2).

The deeper holes, where some of the mortar was also lost, were completely filled with mortar during the 1842 restoration. The grouting work was more extensive than what was needed to fill the holes, perhaps as a way to prevent the surrounding tesserae from falling off (Figure 4.5.3).

The grouting work was done in gypsum and was done quickly and carelessly. Some of the grout was painted on the surface with a dark-brown neutral paint. In some places, the grout was painted to imitate the colours of the surrounding tesserae, while in other places, the grout was painted to replicate the missing part of the mosaic image (Figure 4.5.4).

The surfaces were hard to make out, as they were covered by a thick brown layer which had formed from the oxidation of different kinds of glues and paints used to revive the mosaic. There was also a thick layer of soot from candle smoke on the surfaces (Figure 4.5.5).

Figure 4.5.2 Tesserae fallen off reveal the coloured backgrounds of setting bed.

Figure 4.5.3 The grouting covers a larger mosaic surface than the gap.

Figure 4.5.4 South wall. Detail: the painted plastering.

Figure 4.5.5 North wall. Mosaic before the cleaning, detail.

In several areas, the presence of saline efflorescence caused alterations, specifically the deterioration of the surrounding glass tesserae.

In two fragments in the north transept there was a considerable detachment of the bedding mortar from the wall surface and disintegration of the supporting mortar. The entire wall had been subject to water infiltration from outside the church, and the consequent dampening and drying of the surface had formed soluble salts and the subsequent disintegration of the constitutive materials. This phenomenon had a particular effect on the limestone tiles used for the Apostles' garments which, in fact, were greatly deteriorated (Figure 4.5.6). Defects in

Figure 4.5.6 North transept, degradation limestone stone tesserae.

the adhesion of the tesserae to the setting bed are minor. Some glass tesserae have lost their colour and have become fragile on the surface. The gold-leaf tesserae had many adhesion defects as well as micro-fractures in the cartellina.

4.6   Interventions

S. Sarmati, N. Santopuoli and E. Concina

The intention behind the whole restoration process, from the planning stages to its implementation, was to carry out interventions centred on conservation and maintenance while applying the rules of critical restoration and the principle of minimum intervention to preserve the already well-established image of the basilica after the 1842 restoration.

The few fragments that remained from the extensive mosaic decoration that embellished the church had been "framed" within the new plastered surface. The conservative restoration was performed in accordance with a methodology aimed at examining the documentary aspect of the work while giving also a new visual unity to the many, seemingly unrelated fragments so that the mosaic cycle could be seen or at least imagined in its entirety.

Before starting any intervention on the mosaics, a thorough examination of the wall plasters was made with thermography to see if any other mosaic fragments had been covered during the 1842 restoration.

Thermographic analysis is a very useful diagnostic tool. It is based on the principles of Thermodynamics according to which each body is identified by its thermal emission as a function of its surface temperature which in turn is conditioned by the material's thermal conductivity and specific heat. This non-destructive analysis is able to provide specific answers with a degree of accuracy that varies according to the context and purpose of the investigation.

In this case, it was decided to perform an "active" analysis whereby the thermal emissions were amplified through induced heating.

The analysis revealed discontinuities in the surfaces below the plaster and allowed us to rediscover the figure of the angel between the third and fourth windows. The mosaic fragment was in excellent condition but had been covered during the 19th-century restoration because of the lack of the upper part of the angel's head (Figure 4.6.1, 4.6.2, 4.6.3).

Before and during the restoration of the roof, the mosaics had been covered with tissue paper to be protected against possible damage during the work. The first step in the restoration process was to remove this tissue paper from the mosaic surface, except where the tesserae were at risk of falling off. The cleaning was performed using a pH-controlled ammonium carbonate solution with calculated contact times. This product was used to remove dark soot deposits and deposits from old protective substances, which had, by that point, oxidised. The cleaning process was completed by mechanically removing dirt residue from the mortar in the interstices between the tesserae. A Japanese paper with deionised water was applied to remove any residual salt.

Various portions of the mosaic that had been completely covered were brought back to light by removing the old plasters, which was done only after covering the tesserae around the edges with tissue glazed to prevent them from falling off.

By removing the layers of plaster that covered the edges of the mosaic fragments, it was possible to recover a large part of the original frames, including those around the windows.

Figure 4.6.1 Nave. The third angel under the plaster.

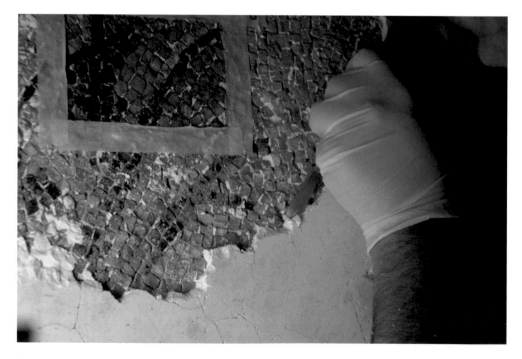

Figure 4.6.2 Nave. Removing the plaster on the third angel.

Figure 4.6.3 The third angel after restoration.

Figure 4.6.4 The bullet inside the gap.

While holes in the plaster were being eliminated, some antique bullets were found in some of the deeper holes. In the 16th century, in fact, Turkish soldiers used to remove lead from the roof to make bullets that were shot sometimes at the mosaics for fun[16] (Figure 4.6.4).

The deteriorated setting mortars were consolidated by infiltration with ethyl silicates. For the consolidation of the dangerously detached base layers of mortar, a pre-mixed hydraulic lime mortar was used. Threaded pins with pieces of canvas on top were inserted into the deeper holes and fixed to the surface of the plaster with acrylic resin.

Tesserae that were not well adhered to the substrate were reattached with hydraulic lime mortar.

During the restoration of the gold backgrounds, numerous *cartellinas* that were either partially detached or micro-fractured were carefully and accurately secured, as well as the metal foil that had been left exposed after the total or partial loss of the cartellina itself. This was done by ethyl silicate infiltrations or treatments (Figure 4.6.5).

The areas where the tesserae had fallen off but the setting bed and coloured backgrounds were still intact were left exposed, and the abrasions in the mortar were touched up with watercolours.

In the areas where the holes were rather deep and the original setting bed had been lost, the missing tessellation was reconstructed with a mixture of lime and marble dust. The mortar was modelled using moulds to replicate the size and pattern of the surrounding tesserae and to be in keeping with the textured surface of the original mosaic. The fresco technique

Figure 4.6.5 The consolidation with ethyl silicate infiltrations.

allowed them to match the colour of the mortar to that of the surrounding tesserae. The reconstructed elements are generally small holes either on gold backgrounds or with repetitive geometric patterns.

The gaps present in the faces were small and easy to reintegrate. The largest gap was located on the face of the third angel, which had been found underneath the plaster. The missing parts, including the top of the head, the hair and a part of the aureole, were reintegrated with moulded mortar so as to restore the visual unity of the figure (Figure 4.6.6, 4.6.7).

Apart from aesthetic reasons, integrating gaps in the mosaic is often necessary in order to ensure better conservation of the mosaic surface. Discontinuity in wall mosaics, when tesserae jut out or hang at different angles, increases the precariousness of the contiguous areas as the free-hanging edges become points of extreme fragility.

This restoration, which was necessary for the preservation of such historically and artistically significant mosaics, was able to recover the incredible colours of the mosaic surface by using a large amount of colours and materials.

Now all the details that were once covered by the dark deposits accumulated over time can be admired again in all their splendour.

Moreover, on the north wall, which has most of the mosaic fragments, a mortar layer, added at the *arriccio* level, gives continuity to the decorative scheme, divided into horizontal bands by elaborate and refined frames. That allows us to perceive and enjoy again the original figurative scheme.

Figure 4.6.6 Retouching the plaster.

Figure 4.6.7 During the retouching.

Notes

1 For a detailed description and interpretation, with complete bibliography, cf. M. Bacci, *The Mystic Cave: A History of the Nativity Church in Bethlehem*, Brno, Rome, 2017, pp. 136–198.
2 M. Bacci, *The Mystic Cave*, Viella, March 2017, p. 136.
3 Bacci, *The Mystic Cave*, p. 123.
4 Our supposition finds little evidence in source documents, but the fact that the bottom of the mosaic stops right at the wooden architrave and that there are 30 cm of wall between the upper cornice/ frame of the mosaic and the roof trusses makes us presume the existence of a ceiling, which probably was decorated with gilding like the architrave. Surely, there had to be a wooden ceiling before our mosaics.
5 Bacci, *The Mystic Cave*, p. 187.
6 The setting bed was applied on a base layer of plaster working on a scaffold (*ponte*). The layer of plaster was laid down in very large patches corresponding to the height and length of the plan of scaffolding. Hence these long wide bands are called *pontate*.
7 *Il mosaico parietale*, P. Pogliani & C. Seccaromi (eds.), Nardini Editore, Florence, 2010, p. 26.
8 Surface area onto which the final plaster layer (setting bed) is applied, so that all tesserae can be embedded while the mortar is still damp.
9 Technique in which the mosaic is illuminated from one side only. Raking light is used to reveal a mosaic surface texture.
10 In gold and silver tesserae, the very thin glass film (0,81 mm) applied onto the metal leaf as a protection.
11 Verità.
12 Texture, in the context of mosaic, refers to qualities of surface which depend on the type of materials used and their cutting, on the direction of the rows of tesserae, and on the width of interstices (M. Farneti).
13 Lucy-Anne Hunt, others.
14 TGA made it possible to estimate the calcium carbonate content on the basis of the weight loss occurring between 600 and 800°C, whereas the organic matter can be determined from the weight loss in the range 200–500°C. The decomposition of calcium hydroxide takes place in the same range.
15 M. Bacci, Old restorations and new discoveries in the Nativity Church, Bethlehem. *Convivium*, 2(2) (2015), 36–59.
16 Bacci, Old restorations and new discoveries in the Nativity Church, Bethlehem, p. 46.

4.7   Floor mosaics – archaeological excavations

G.A. Fichera

4.7.1   Introduction

The first and almost unique archaeological excavation carried out within the Nativity church in Bethlehem dates back to 1934 and was led by archaeologists from the Department of Antiquities of the British Mandate in Palestine.

The archaeology of those years could not rely on today's advanced methodologies of investigation, and excavations were not functional to research projects but focused on restoration or consolidation. In particular, following the destructive earthquake in the Bethlehem area in 1927, the Palestinian government commissioned William Harvey to carry out the structural analysis of the basilica to verify any disruption. For an exhaustive summary of the interventions made in those years the reader can refer to the bibliography in Bacci *et al.* (2012) and Bianchi *et al.* (2016). The British excavations brought to light an exceptional mosaic decoration of the floor of an older basilica, dating back to the time of Emperor Constantine (272–337) and hidden for centuries under the floor of Justinian's church. At that time an accurate planimetric survey of that mosaic decoration was carried out.

Ever since the theories that have accompanied the historical and architectural interpretation of one of the most important monuments of Christianity have been multifarious and different, in particular in relation to the correct chronological attribution of the two buildings, constantly in doubt between the Constantine and Justinian phases. The analytical stratigraphic interpretation of the architecture (Bacci *et al.*, 2012) and the dating of the wooden architraves placed above the columns of the aisles (Bernabei and Bontadi, 2012), both completed during the 2010 diagnostic campaign, and the recent study of Bacci (Bacci, 2017: 59–68) where he proposed a thorough survey of extant studies and supported the archaeological evidence with further arguments from sources, solved the problem by attributing the current building to a great intervention of integral reconstruction dating back to the age of Justinian (482–565).

Recently, during the restoration works in the church, the possibility was offered to carry out new and focused archaeological excavations in view of a more detailed and analytical knowledge of the historical events that occurred from the construction and destruction of the first basilica up to its total reconstruction. The new data collected can be useful for discovering further evolutionary phases or shedding light on some remaining obscure aspects in the history of the building.

The new restoration projects of the floor mosaics, already visible since the 1930s and covered with wooden trapdoors, enabled a more correct understanding of their structural context. The archaeological excavation brought to light new elements useful for the historical reconstruction of the vicissitudes undergone by the basilica over the centuries.

Between 2018 and the beginning of 2019, five different sectors, corresponding to the mosaic areas shown in Figure 4.7.1, were opened. Below are the preliminary results of a project that has not yet come to an end. The archaeological data and the findings are still under study.

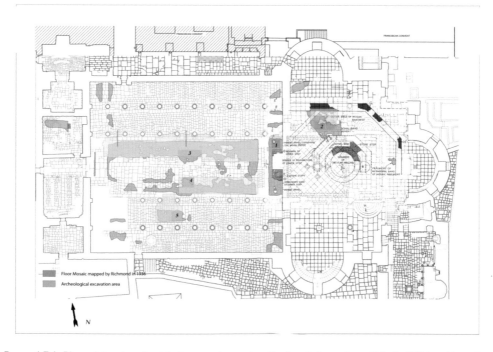

Figure 4.7.1 Plan of the church with the mosaics and walls already brought to light in 1934 and location of the recent archaeological surveys.

Figure 4.7.2 Area 1 during excavation. View from the east.

4.7.2   Area 1

Area 1 is located in the eastern edge of the central nave, in direct contact with the presbytery wall. It measures approximately 2 m × 2 m and marks a portion of the mosaic floor characterised by a geometric meander decoration with some letters forming the Greek word for fish, ΙΧΘΥΣ, (anagram meaning "Jesus Christ Son of God Saviour"), framed by a wide white frame (Figure 4.7.2).

The archaeological excavation has verified how the area had been entirely excavated in the 1930s and subsequently filled with earth and stones that contained numerous white mosaic tiles coming from the partial destruction of the mosaic floor.

The architectural structures delimiting the surveyed area are extremely representative of the main construction phases and transformation of the two basilican buildings, even if the original stratigraphic layers already excavated by the British are missing.

On the eastern side, a mosaic floor and a wall belonging to the Constantine basilica were brought to light. The wall, according to the planimetric reconstruction proposed by Richmond (Richmond, 1936, Plate XXXVI), correspond to the western perimeter of the octagonal structure that formed the original east side of the basilica. The wall was covered with a layer of white plaster without further decorations. At -0.76 m from the current floor is the mosaic floor of the first basilica, subsequently destroyed to allow the construction of the foundations of Justinian's columns. In the southern slope the excavations brought to light the wall structures that defined the original system of access to the grotto from the centre of the nave. It is possible to see the entrance in the photos taken in 1934 during the excavations that involved the entire nave of the basilica (Figure 4.7.3).

The foundation (stylobate), on which the columns of the basilica of the Justinian age were built, was rediscovered on the northern side. On both bases of the columns visible in the survey some red lines were identified, two of which are horizontal and placed at the same height, whereas a third one is vertical and marks the position of the column base (Figure 4.7.4). The discovery of similar signs placed at the same height on the basements of the western columns led to the hypothesis that such signs may have been related to the original design of the building in the Justinian age. In particular, the lines were used to define

Figure 4.7.3 Picture from 1934 with the old access system to the grotto.

Figure 4.7.4a Area 1. a: Construction lines in red on the base of the N-W column.

Figure 4.7.4b Area I.b: Construction lines in red on the base of the N-E column.

the exact positioning of the columns along the stylobate and the height of the floor of the basilica. This is an exceptional discovery due to the rarity of this kind of traces in sixth- and seventh-century monuments. The lines provide answers also as to the hitherto unanswered question concerning the height of the Justinianic floor. The stylobate, made of large ashlars perfectly squared with the finished surface, was built within a foundation trench that destroyed a large portion of the mosaic floor.

Finally, in the west side of the survey the excavation revealed a very rough wall structure made of a row of large stones without mortar and directly set on the mosaic floor, for this reason partially removed. As can be seen from the photographic images published by Richmond in 1936 (Figure 4.7.3), the structure crosses the entire nave in a north–south direction.

4.7.3   Area 2

The second survey is located in the northern apse, between the steps to the raised presbytery area, and measures 4.36 m × 3.65 m.

The present-day floor is made of square-shaped marble tiles with dimensions ranging from 50 × 50 cm to 70 × 70 cm, while the nave floor is composed of stone tiles. The mosaics already visible before the archaeological excavation were at a lower depth of about 30 cm. This indicates that also the presbytery area of the first basilica had a higher elevation than the nave.

The removal of the marble tiles enabled us to delimit and dig the filling, dating from the 17th and 19th century interventions. The layer was composed of earth and stones and covered the northern wall of the Constantine church already present in Richmond's plans, aligned on an east–west direction, and part of the mosaic floor that leaned against it. The perimetral wall, deliberately destroyed at a perfectly horizontal height, was characterised by the presence of some irregularly shaped ashlars, some of which defined a relieving arch, probably linked to the presence of cavities in the subsoil (St Jerome's grottos). One wall was also discovered, aligned on a north-east–south-west direction, consisting of a single facing of ashlars and a mosaic floor both to the east and to the west. These elements can be interpreted as belonging to a dividing wall or a step inside the octagon but not as a perimetral wall with a supporting function (Figure 4.7.5).

The white mosaic floor presented in two points evident traces of an ancient restoration, probably due to prolonged consumption, through the positioning of new tesserae that did not respect the general texture and the installation height.

Although the area was entirely marked as excavated in the British drawings, two intact stratigraphic layers were identified in the west and east parts. This new discovery enabled us to reconstruct the plausible sequence of the events that led to the destruction of the first basilica and the reconstruction of the new building. These layers are composed essentially of lime mortar, which can be interpreted as preparation levels for the above floor. Furthermore, in the eastern edge of the area one of these levels covered a patch of decorated mosaic whose existence was unknown. The mosaic decoration is composed by two coloured braids placed to delimit a hexagonal medallion housing a pomegranate tree with four fruits hanging from the branches. This pattern integrates with the decorative cycle of the northern apse with plant motifs and animals (Figure 4.7.6).

Figure 4.7.5a Area 2. a: Wall of the Constantine church from the south. At the centre of the wall there is a circular structure that can be interpreted as a relieving arch.

Figure 4.7.5b Area 2. b: Area during excavation. View from the west.

Figure 4.7.6a Area 2. a: Mosaics after restoration. View from the south-west.

Figure 4.7.6b Area 2.b: Detail of the pomegranate tree brought to light during new excavation.

Apart from bringing to light this exceptional discovery, the analysis of stratigraphic layers sheds light on the historical events associated with the reconstruction of Justinian's basilica by contradicting the narrative according to which the first building would have been totally destroyed by fire. The data emerging from the analysis clearly indicate that Justinian's building was implanted directly on the mosaic floor of the previous church.

4.7.4 Area 3

The third survey corresponds to an area 23.5 m long and 2.6 m wide, located on the northern side of the nave.

The removal of the stone tiles and of the mortar level used as support for the latter enabled us to identify a large portion of the original stratigraphy, left in place from the excavations of

Figure 4.7.7 Picture of the 1930s with the nave view from the west. In black, the unexcavated stratigraphy.

the last century. These levels developed along a band of about 40 cm parallel to the columns of the nave, as can also be seen from old photographs (Figure 4.7.7). This is a very important deposit that allows us to know in detail all the construction phases of the sixth-century basilica and its subsequent transformations.

The analysis of the entire northern section was followed by two surveys, the first of which is located between the fourth and seventh columns from west to east (Figure 4.7.8).

Even in this case the accurate analysis of the mosaic floor led to the identification of a series of restorations that are characterised by substantial differences. This variety suggests a long life of the Constantine basilica that caused the consumption of the floor and the need to restore it, in a period certainly older than the reconstruction works of Justinian's era (Figure 4.7.9). In some cases, for example, SU 338 and SU 340, interventions were carried out by laying a mortar layer to fill the gaps that had occurred in the mosaic floor to recover a homogeneous trampling level. In other cases, such as SU 336, the gaps were filled by using well-laid stone mosaic tesserae, which were characterised by a different orientation in the texture or by a colour being distinct from the white chromatic base. A third type of historical restoration had instead used mortar mixed with mosaic tiles, perhaps to fill a deeper gap.

The foundations of the new building were placed directly on this mosaic floor in Justinian's age. In fact, we investigated layers made of sand and lime lenses and composed almost exclusively of stone flakes; these are all elements that well characterise the different

Figure 4.7.8 Area 3 during excavation. View from the east.

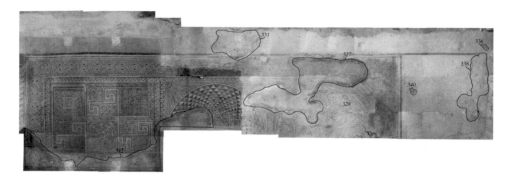

Figure 4.7.9 Area 3. Photo-plan of the floor mosaic with different stratigraphic units.

Figure 4.7.10 Area 3. Traces of white plaster on the internal facade of the Constantine church view from the east.

processing phases taking place in the construction site. Moreover, near the façade wall of the Constantine church, more backward of about 2 m compared to the current one, the excavations discovered evident traces of the definitive destruction of the old basilica, occurred in concomitance with the construction of the new building. In the layers of construction was found a very high percentage of fragments of coloured plaster belonging to the original decorative apparatus of the walls. In the only row of preserved masonry a small fragment of white fresco is preserved (Figure 4.7.10).

In a later phase, a wide trench parallel to the current stylobate, distant from it between 0.2 and 0.5 m and 1.6 m deep, was excavated on the surface of these layers. A band of white mosaic tiles lying parallel to and spaced 30 cm from the wall at the latter's edge is preserved on the points where the pit was narrower. In Area 1, we found the same frame leaned on the walls and dating from the Constantine age. This element has allowed the hypothesis that the oldest wall should be exactly in the same position as the new one, reconstructed as a base for the new column system. The excavation served in the past to completely remove the old colonnade to rebuild the new one by maintaining the same position of the previous one (Figure 4.7.11).

Once the reconstruction of the new stylobate was completed, the foundation trench was filled. A number of glass objects were placed at the contact point between the façade of the old destroyed church and the base of the first column. It can be assumed that they belonged to at least two glass lamps, as is suggested by the presence of two different types of handles. The site of the discovery, corresponding to the point where the old façade had been destroyed for the construction of the new colonnade, and more specifically the surface of the filling itself, suggests a precise intention to perform a foundation ritual (Figure 4.7.12). The removal of glass findings was a very delicate operation that led to the identification of at least four different levels of overlapping fragments (Figure 4.7.13).

Figure 4.7.11 Area 3. Foundation pit for the columns of the Justinian church. View from the south-east.

Figure 4.7.12 Area 3. Foundation fill after the built of the stylobate. View from the south.

The different layers of Justinian's construction site were covered, at the end of the works, by a well-levelled lime level and by a subsequent stratus of very pressed sand mixed with stones. These layers were functional to create the preparation screed of the floor, the trace of which has not been preserved but whose existence is attested by the presence of red lines on

Figure 4.7.13 Area 3. Detail of the glass lamps found in the foundation fill.

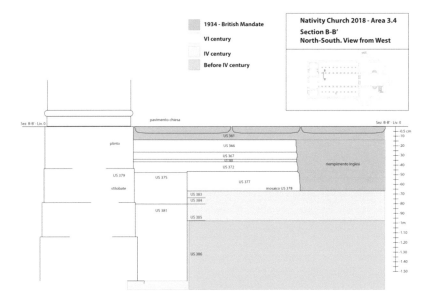

Figure 4.7.14 Section of Area 3 representing the stratigraphy in the nave and the chronological periods.

the masonry face of the stylobate. The latter can be interpreted as design signs functional to the correct positioning of the columns and the floor plan (Figure 4.7.14).

The subsequent stratigraphy refers, in the central portion of the nave, to at least two new floor levels composed, the first, by tiles placed in a dark-grey mortar. The latter's presence

Figure 4.7.15 Area 3. Floor level in tiles and grey mortar. View from the east.

confirms, in combination with the numerous fragments of tiles found in the different layers, that the roofs of both churches were made of tiles (Figure 4.7.15).

The second floor level consisted of a layer of *cocciopesto* preserved only in small fragments at the contact point between the base and the stem of the columns (Bacci, 2017: 239). The *cocciopesto* layer must have been removed later, breaking the base of the columns in some places, perhaps to build the last floor made of stone tiles (Figure 4.7.16).

Figure 4.7.16 Area 3. Red lines at the base of the column and the remains of the *cocciopesto* floor.

4.7.5   Area 4

The fourth survey corresponds to an area of 12.20 m long × 2.5 m wide located in the southern side of the central nave, delimited in the western edge by one of the wooden trapdoors left after the 20th-century archaeological excavations.

The survey area had already been entirely excavated in the last century, as can be seen from the graphic documentation. The pictures showed the completely exposed floor mosaic and, similarly to what was observed in the northern side, a large portion of original stratigraphy developed along a strip about 40 cm parallel to the columns of the nave (Figure 4.7.17). The documentation of the archaeological section and the comparison with the opposite side of the nave confirm a substantial stratigraphic analogy. The construction site of the Justinianic era was implanted directly on the mosaic floor and, after the accumulation of layers essentially composed of stone processing waste and from lime lenses left in place, it is sealed with a level of mortar. This last level of preparation for the floor is carefully prepared, but even in this case, no trace of it was found, as it was removed in ancient times.

Within the limits of this survey, in a large gap on the mosaic floor created during the reconstruction of the Justinianic church, it was possible to carry out a small survey (2.5 m × 0.5 m) to verify the oldest stratigraphic sequence on which the construction site of the Constantine church had been established (Figure 4.7.18).

As already verified in the sections of the foundation trenches of the stylobate, below the mosaic, which is located at about -0.70 m from the current floor of the nave, a succession of layers was dug associated with the mosaic preparation. These layers were composed of

Figure 4.7.17 Area 4. Visible under the columns the unexcavated stratigraphy.

Figure 4.7.18 Area 4. Gap in the mosaic floor where was carried out an excavation survey.

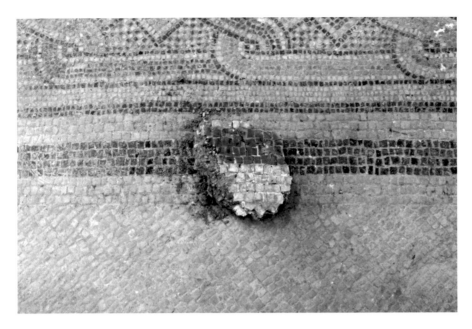

Figure 4.7.19 Area 4. Fragment of mosaic found in the survey in comparison with the mosaic in place.

stones on which more and more refined levels of mortar had been spread up to the bedding for the mosaic tiles, characterised by the presence of ash and charcoal in the cement mixture that give to the mortar a characteristic dark-grey colour.

At a depth of approximately -1.36 m below the current floor, the excavations brought to light layers with some fragments of a mosaic floor in black and white tesserae, to which the preparation layer was still attached. The discovery is in itself exceptional, even if the floor level discovered is in a fragmentary state and cannot be considered to be in its original placement, since it clearly shows that the construction site of the Constantine church was placed on an area already affected by the presence of wall structures associated with exquisitely crafted mosaics. It is also necessary to specify that the newly discovered fragments are characterised by a black and white bichromy, even if one of the larger fragments showed a band of black tesserae composed of four rows, adjacent to a band of white tesserae ranged in four rows, and from a last row of tiles laid diagonally (Figure 4.7.19).

The overlap with the floor mosaic in place showed a total analogy of dimensions and decorative motifs. However, since the mosaics were found not only at a lower depth of more than 0.7 m but above all that they were perfectly sealed by the level of preparation of the mosaic floor in place, it can be ruled out that the fragments found in the survey may belong to the floor level of the Constantine age. It could be floor levels belonging to structures or buildings already present in the area of the basilica but not too ancient and therefore characterised by the presence of similar decorative motifs, or even a first realisation then destroyed and replaced by a new floor level at a highest share. Still unknown are

Figure 4.7.20 Area **4**. Natural rock at the bottom of the survey. View from the east.

the reasons that led to the destruction of the first mosaic floor and to re-use the fragments as a level of preparation of the construction site, in direct contact with the local natural rock bed that is located at a height of -1.55 m below the current stone floor of the nave (Figure 4.7.20).

The stratigraphy is intact, and the findings seem to refer to a substantially homogeneous context. These are results of considerable importance given the small size of the survey carried out and the preliminary status of the studies presented here. The continuation and deepening of the investigations will bring new data to an absolutely extraordinary site that keeps revealing incredible surprises and never ceases to amaze archaeologists, restorers and the thousands of visitors who have been visiting it for centuries.

Figure 4.7.21 Area 5 during excavation view from the south.

4.7.6 Area 5

The fifth survey corresponds to an area of dimensions equal to 4.20 m × 4.40 m located in the south aisle.

Below the stone tiles and the level of mortar used to lay them in, the level of well-laid stones was uncovered in a layer of sand already identified in the northern side of the nave and interpreted as the level of preparation of the roof of the Justinian age. The difference consisted in the preparation screed that did not go beyond the space between the columns of the central nave in the only case in which the archaeological excavation concerned one of the aisles.

The layer of mortar that permanently sealed the building layers with two small holes on the surface was later identified below this level. In the filling of one of the holes were found fragments of glass and among these also two small handles. In consequence of this important discovery, the hole was therefore interpreted as a small "sacred pit" in which liturgical objects found during the reconstruction of the church in the Justinianic age were concealed (Figure 4.7.21).

The foundation cut of the columns had been made on the surface of a layer of gravel and stone flakes of slight thickness that directly covered the mosaic floor. The excavation of the filling cut allowed us to verify that, at a depth of -1.5 m below the level of the current floor, the cut had reached the natural rock bed. After the total removal of the older wall, it was carefully worked and flattened with a tool whose working marks remain evident, in order to create a perfectly horizontal surface for the new wall (Figure 4.7.22).

Figure 4.7.22 Area 5 at the end of the excavation. View from the east.

This is the only point in the basilica where it was possible, due to the size of the excavated area, to reach the bottom of the foundation trench with the perimetral walls of the basilica of the Constantine age.

4.7.7   Summary of the main periods identified during the excavations

4.7.7.1   Period I (before the fourth century)

The stratigraphy identified in section and in the small survey carried out in the nave below the mosaic floor belongs to a more ancient period than the construction of the first basilica, and therefore earlier than the fourth century. These are layers of stones and preparation levels for the mosaic floor. At the base of the stratigraphy, in direct contact with the natural rock bank, some fragments of a broken mosaic were used as levelling material, which could belong to a pre-existing building or to a first draft of the mosaic decoration of the Constantine basilica.

4.7.7.2   Period II (fourth century)

In relation to the Constantine basilica, some of the squared blocks are carried out with a grey mortar mixed with ash (similar to the layers of preparation of the mosaic floor) that defined the original façade, more backward than almost 2 meters compared to the current one, already detected at the time of the British excavation. However, the excavation of the foundation filling trench for the reconstruction of the base of the columns allowed us to identify the foundation row of the façade itself, of which a large block has been preserved inside the same trench. This fact allows us to assume with certainty that the previous foundation wall of the columns was in exactly the same position and was completely disassembled for later reconstruction.

The interior façades of the walls were also covered with decorated frescoes that were destroyed, with the wall itself, shortly before the excavation work began for the construction of the new stylobate. The Justinianic site was therefore placed directly on the mosaic floor, occupied by a thick layer of sand mixed with the remains of the fresco destroyed, on which the trenches were subsequently excavated for the construction of the new walls.

4.7.7.3   Period III (before the sixth century)

The life span of the basilica has been further extended thanks to the identification of at least three different types of ancient restoration of the mosaic. Interventions that were evidently carried out before the integral reconstruction to fill some gaps that had formed in the mosaic floor.

4.7.7.4   Period IV (sixth–seventh centuries)

In relation to the Justinianic church, the recent archaeological excavations brought to light a very clear situation, showing that the first operations of the new construction site led to a thorough destruction of the walls of the previous building, whose façade was decorated with frescoed plaster. On the layer made essentially of sand and plaster fragments, the foundation trenches were dug for the construction of the new stylobate. This base, composed of three rows of square elements but not finished on the surface, completely replaced the previous one, which however occupied exactly the same position.

During the building works and therefore as the construction site was still in full operation, the layers of mortar and flakes of block processing can be interpreted as evident signs of the different processing phases that took place at the base of the newly built columns. When the foundation trench had been filled, some glass objects, perhaps lamps, were concealed with an evident ritual purpose in the very point where the base superimposed the corner of the original façade. A second "sacred pit" containing remains of liturgical objects was identified in the south aisle.

The layers of workmanship, probably when the construction site had been completed, were sealed by a plane of mortar and by a level in very pressed earth mixed with stones, functional to place the floor of which no trace has been preserved.

The construction continued with the laying of the plinths for the columns, which were decorated with signs of design including a horizontal line of red colour that marked the level of the floor and small vertical lines that marked the position of the column above. This proves to be a very important testimony, given the rarity of such traces.

In the centre of the nave, above the preparation layer, traces of two further floor levels remain, following the second basilica, one in dark-grey mortar with tiles and one, in *cocciopesto*, already removed in ancient times, also breaking the base of the columns.

4.7.7.5   Period V (1930s)

British excavation and construction of the walls that defined the mosaics.

References

Bacci, M. (2017) *The Mystic Cave: A History of the Nativity Church in Bethlehem*. Viella Editore: Brno, Rome.

Bacci, M., Bianchi, G., Campana, S. & Fichera, G. (2012) Historical and archaeological analysis of the Church of the Nativity, in the Church of the Nativity in Bethlehem: An interdisciplinary approach to a knowledge-based restoration. *Journal of Cultural Heritage*, 13, 5–26.

Bernabei, M. & Bontadi, J. (2012) Dendrochronological analysis of the timber structure of the Church of the Nativity in Bethlehem, in the Church of the Nativity in Bethlehem: An interdisciplinary approach to a knowledge-based restoration. *Journal of Cultural Heritage*, 13, 54–60.

Bianchi, G., Campana, S. & Fichera, G. (2016) Archeologia dell'Architettura nella Basilica della Natività a Betlemme, pp. 1567–1589, in Olof Brandt – Vincenzo Fiocchi Nicolai (a cura di) Acta XVI Congressus Internationalis Archaelogiae Christianae Romae (22–28.9.2013). *Costantino e i Costantinidi. L'innovazione costantiniana, le sue radici e i suoi sviluppi*. Pontificio Istituto di Archeologia Cristiana, Città del Vaticano.

Richmond, E.T. (1936) Basilica of the nativity: Discovery of the remains of an earlier church. *QDAP*, 5, 75–81; 6 (1937), 63–72.

Chapter 5

Plasters

S. Sarmati And N. Santopuoli

5.1 Material composition

The basilica's internal walls were originally covered with marble slabs, used for hiding the masonry structure with a valuable decorative material. The slabs were fixed to the wall with a layer of mortar and with bronze hooks, which in some cases are still *in situ*, fixed into holes in the stone blocks.

The holes are still visible on the basilica's walls, in the areas where the masonry was left exposed, such as the narthex area. Some of the holes still have inside the bronze hooks and the Proconnesian marble pieces used to fix the slabs.

It is very likely that the material of the covering slabs and the material of the stone pieces found within the holes is the same.

The final effect was probably a monochromatic white surface with cerulean shades and grey veins.

During the restoration in the southern apse, next to the door, black marble pieces were found, testifying to the presence, in the past, of a dark frame around the doorframe.

The marble slabs covered from the bottom to the top the internal surfaces of the side walls of the aisles and the wall areas above the architraves of the internal colonnades in the aisles.

Marble slabs surely covered the lower part of the counter-façade, the choir walls and the apses walls too.

The restoration work was carried out according to the principle of the critical restoration and the minimum intervention, in order to maintain the consolidated image of the basilica after the latest important restoration, made in 1842.

After centuries of neglect, the Greeks, during the patriarchate of Athanasios III, with the approval of the Ottoman Sultan Mohamed Ali, started the restoration works of the basilica, with the renovation of the roof. The walls, almost bare, were covered with white translucent plaster, and the few fragments of the large wall decoration which embellished the church were surrounded by the new plaster.

The white plaster covered all the walls of the church, with black linear frames surrounding the windows. The perimeter of the three apses was surrounded by simple frames with floral decorations. A 1-metre-high dark-grey flounce covered the lower part of all the walls of the basilica.

The plaster on the walls was applied in different layers. The intermediate layer was composed of a compact coating with a large amount of vegetable fibres. There was a topcoat on it. Decorations were made in a second phase, with a black pigment mixed with lime water.

Before the restoration, samples of plaster were taken from significant areas of the wall in order to identify the composition of the mortars.[1]

The samples were taken in the upper part of the walls in order to avoid the risk of a sampling made from areas where a restoration already had been made (Figures 5.1.1, 5.1.2).

Figure 5.1.1 Sample 1. The sampling area.

Figure 5.1.2 Sample 2. The sampling area.

Thermo-gravimetric analysis (TGA) and Raman spectroscopy analysis for the qualitative characterisation of the pigment, and Fourier-transform-infrared-spectroscopy (FTIR) for the characterisation of binders were carried out on the samples.

The analysis detected the presence of lime, used as a binding agent, and carbonate material and very fine sand, used as aggregate.

The decorations' colour is a black carbon mixed with lime. The same pigment was used in small quantities in the superficial mortar.

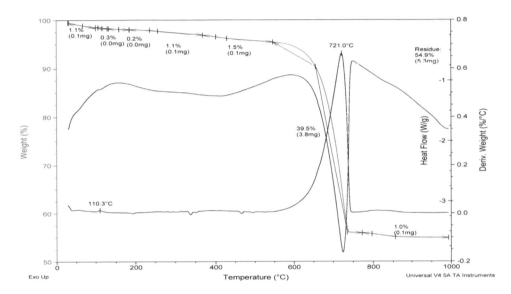

Figure 5.1.3 Sample 1 analysis TGA-DSC.

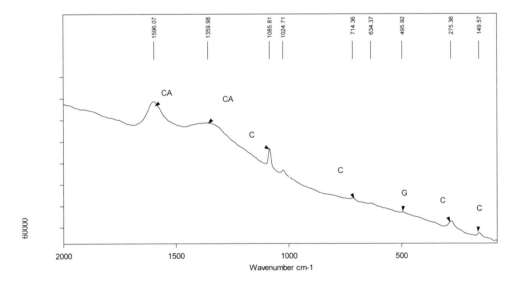

Figure 5.1.4 Raman-spectroscopic-analysis. CA = coal, C = calcite, G = gypsum.

Figure 5.1.5 Sample 2, detail.

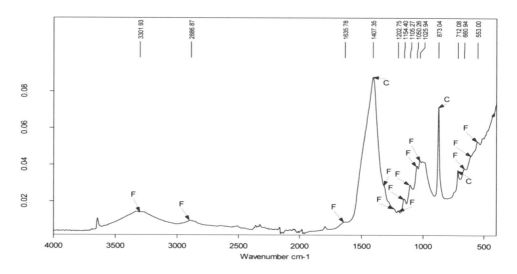

Figure 5.1.6 Sample 2 infrared spectroscopic analysis (FTIR); C = calcite. F= Hemp fibers

In one of the intermediate layers there is a large amount of hemp fibres mixed with lime mortar and carbonate aggregates.

The analysis carried out on the superficial layer of the plaster detected the presence of a light-grey superficial plastering mixed with a black carbon pigment used to create a light-grey gradation on the surface.

In conclusion, mineralogical analysis, carried out by the *Centro di Ricerca Scienza e Tecnica per la Conservazione del Patrimonio Storico-Architettonico*, of the Department of Engineering of the Sapienza University in Rome, detected the presence of lime-based plasters, generally applied in two or more layers. The deeper layers are composed by lime mortar

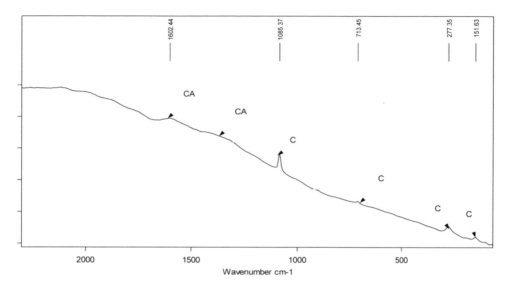

Figure 5.1.7 CA = coal C = calcite.

and hemp fibres, and the superficial layer is composed by lime mortar and a protein organic binder, on which a thin lime-based veiling had been applied.

5.2 State of conservation

The restoration of the basilica plasters was preceded by an analysis of the present condition of the church and an investigation of the historical events which contribute to its actual appearance.

Some soundings were carried out as preparatory activities, such as the thermographic analyses of all the surfaces and the stratigraphic investigation of the plasters of the basilica.

The plasters were covered by thick deposits of dust, caused by candle smoke, and by dark drippings and dark-brown stains caused by deposits of dusty material and tannin, transported by rainwater infiltrations through the wooden covering (Figure 5.2.1).

There were large plaster losses in the lower parts of the walls. In the areas where the plaster still remained, it lost its adherence to the wall and was raised up (Figure 5.2.2).

The walls of the counter-façade and of the three apses were covered with large deposits of soluble salts, which caused the loss of adhesion and the pulverisation of the plaster (Figure 5.2.3).

Because of the outer walls' degradation, rainwater penetrated inside the basilica, and the repeated cycles of alternate wetting and drying created large deposits of salt efflorescence on the inner plaster's surface.

The entire lower part of the walls had big losses due to the infiltration of water from the ground. These losses were very large on the wall of the right-hand aisle, where the masonry was exposed (Figure 5.2.4).

Figure 5.2.1 The wall of left aisle, detail.

Figure 5.2.2 The wall of the right aisle, detail.

Figure 5.2.3 The wall of the right aisle, pulverisation of the plaster.

Figure 5.2.4 The wall of the right aisle, loss.

Figure 5.2.5 The north apse before restoration.

These lower areas were often made with different kinds of plasters, which had different colours too.

The pictorial decoration around the windows and on the perimeter of the apses was very damaged and dusty and displaycd large lacunae; in the middle apse, the central fastigium had been lost (Figure 5.2.5).

IR thermographic analysis was carried out on all the basilica's surfaces in order to assess the state of conservation of the plasters.

The IR thermography is a non-invasive diagnostic technique which measures the infrared radiation emitted from an object subject to thermal stress and can determine the superficial temperature.

The different elements of an artefact will reach different temperature subjecting to thermal stress, and this allows us to detect not just the different materials but even their state of conservation and identify losses or humidity presence on the plasters.

During the analysis are created maps, in false colours, which represent the investigated areas. This is a realistic representation of the analysed object that shows the variation of the temperature on the surface and is called thermogram.

The instrument used was a thermal imaging camera model Trotec IC080L (Figure 5.2.6). Air heaters were used to heat the surface (Figure 5.2.7).

The thermographic study of the walls not only indicated the degradation of the plasters but also enabled us to investigate further details of the construction of the basilica.

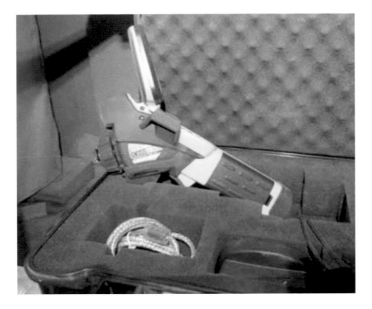

Figure 5.2.6 Thermal camera Trotec IC080LV.

Figure 5.2.7 Air heaters.

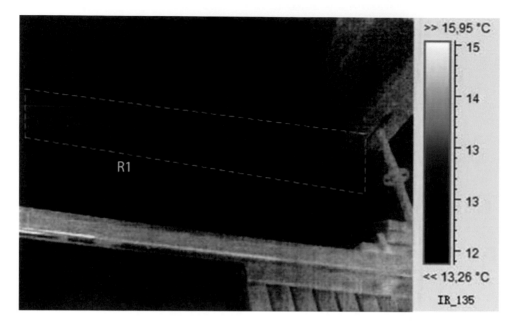

Figure 5.2.8 The thermal image shows a regular hotter area R1 in reference to the wooden sleeper below the plaster lodged in the masonry.

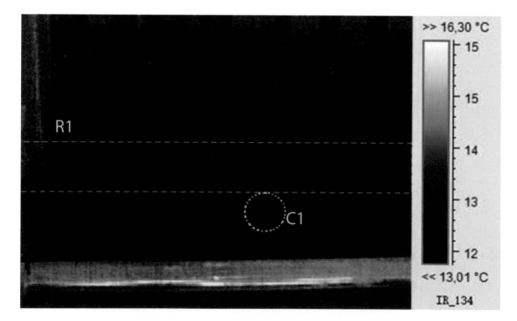

Figure 5.2.9 The thermal image shows a regular hotter area R1 in reference to the wooden sleeper below the plaster lodged in the masonry. The C1 area shows a different kind of material behind the plaster layer it could be grouting or detachment.

5.3   Interventions

The restoration works on the plaster started in June 2014 with the thermographic survey made with the aim to check the presence of underlying mosaics. Then a comprehensive evaluation of the existing plastering layers was made to prepare the mapping of all surfaces and their degradation processes (Figure 5.3.1).

All the plaster surfaces have been consolidated with hydraulic lime infiltrations. The procedure concerned portions of walls whose surface areas displayed cracks or holes. An injection of aqueous and alcohol (50:50) has been applied to such areas in order to remove dust and facilitate the penetration of the hydraulic lime (Figure 5.3.2).

After testing the surface to understand the best cleaning system to remove dirt deposits, it was decided to perform a mechanical cleaning of the surface by removing the dirt with dry cleaning (Wishab sponges) and mechanical tools (Figure 5.3.3).

All improper restorations and cement grouting from previous interventions have been removed.

One of the most important ways to treat salts on the surface is the prevention of water infiltration from the masonry. So after the mechanical removing of the salt deposits and prior to the restoration of the plaster surface, all the external masonry has been restored by removing the improper fillings and introducing new fillings around the stone blocks with the aim to prevent the passage of the rainy water.

Repairs of lacunae were carried out with lime mortar made out of hydraulic lime and dried quartz sand.

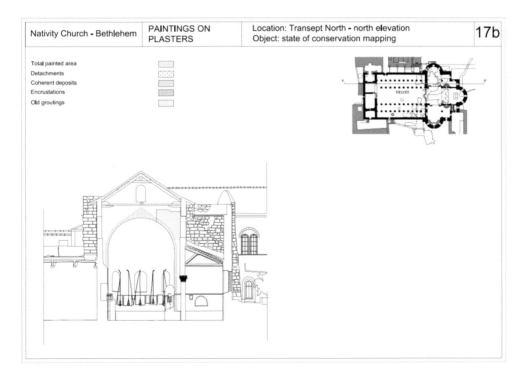

Figure 5.3.1 Table of mapping.

Figure 5.3.2 The plaster consolidating.

Figure 5.3.3 The plaster cleaning.

Figure 5.3.4 The plaster filling.

In the lower areas of the walls where there were moisture problems preference was given to an eco-biocompatible plaster, light dehumidifier: INTOPORE by TCS. In particular, the open porous structure characterising this plaster provides the material with high steam diffusion capacity by regulating the return to the outside of the moisture and thus avoiding the formation of dangerous condensation and bacterial proliferation (Figure 5.3.4).

A light paint, pigmented so as to provide the surface with a homogeneous colour, was applied to the plaster.

All the black frames around the windows were retouched (Figure 5.3.5).

The decorations around the apses were also retouched according to the evidence provided by historical photographs, in the aim to reproduce the parts that had been lost (Figure 5.3.6).

Figure 5.3.5 The frames around the windows.

Figure 5.3.6 The decoration around the apse.

Note

1 *Analysis carried out by the* Centro di Ricerca in Scienza e Tecnica per la Conservazione del Patri-monioStorico-Architettonico (CISTeC).

Chapter 6

Paintings on columns

6.1 The figurative cycle

M. Bacci

In 2015, the restoration works added a new dimension to our understanding of the Crusader reshaping of the Nativity Church through the rediscovery in the transepts of some remnants of wall paintings, whose extension was hitherto deemed to have been limited to the nave and some annexes. Despite their fragmentary state, they bear clear witness to the existence of a wider painted decorum prior to the decoration of the church with mosaics, achieved in 1169. Apparently, the latter was partly superimposed on an earlier program of wall paintings, as is indicated by a very small fragment of painted plaster found below the tesserae on the upper left corner of the *Incredulity* in the north transept. Larger portions of mural paintings were discovered on the east wall of the south transept, close to the present-day Greek Orthodox altar of Saint Nicholas. Despite their fragmentary state, it is still possible to discern the outline of a building with a lunette-shaped façade and an angel represented in three-quarter view and in orant pose on the upper portion of the wall close to the arched window (Figure 6.1.1).

Below this composition, there are some traces of a flaked surface, a visual convention associated in Byzantine art with the representation of military cuirasses (Figure 6.1.2). The presence of soldiers' figures in this area of the church may indicate that they originally belonged to a scene of the *Massacre of the Innocents*, whose burials were worshipped in the underlying caves. Such elements as the chromatic palette, the angel's bodily proportions, and the stylised rendering of the architectural building point to a dating in the first half of the 12th century.[1]

In light of this discovery, it can now be assumed that the Crusaders engaged, since the first decades of their rule in Palestine, in enhancing the basilica with a new painted decorum, even if, shortly later, the latter happened to be mostly substituted with a more precious and lavish one. Sources are not clear enough as to the state of preservation of the church interior when the newcomers arrived on the spot, in 1099. A mosaic cycle had been made there in the late sixth century,[2] and restorations had been made at some point during the first period of Islamic rule, since Patriarch Eutychius of Alexandria (ca. 938) mentions the refurbishment of the south apse (used as a *qibla*) with the display of Cufic inscriptions.[3] The hint in the Russian pilgrim Daniel's travelogue (1106–1108) at the church being decorated with mosaics seems to indicate that at least some portions of the old décors were still to be seen, but possibly only in a fragmentary state.[4]

Figure 6.1.1 Figure in orant pose, fragment of a mural painting on the eastern wall of the south transept.

Be this as it may, an inscription on the fifth column of the inner row of the south aisle, records that the nearby image of the enthroned Virgin "Glykophilousa", displaying the Mother and the Child cheek-to-cheek, was made in the year 1130 (Figure 6.1.3).

The circumstances under which it was painted are clearly revealed by its compositional features: Mary is represented full length, in a lovely attitude vis-à-vis her Son, within a rectangular frame signalling its role as a mural icon. In a marginal, lower position to both sides of the image are represented bowing figures, who can be easily identified as the donors who promoted and financed its public display: to the left, a knight wearing a furred red mantle and a light-brown tunic, kneeling close to a shield (Figure 6.1.4), and to the right, two women whose modest clothing and uncovered hair (Figure 6.1.5) are intended as a way to manifest their devotion and self-commitment to the Mother of God. Each actor of this composition is engaged in a dialogue whose meaning is made clear, on the one hand, by their poses and gestures and, on the other hand, by the Latin verses inscribed on the lower and upper margins of the mural icon. The sense of reading is from the bottom upwards: the donors' request of salvation is made explicit by the nearby inscription *Virgo coelestis confer solatia moestis*, "Heavenly Mother, grant solace to the needy", which finds its counterpart in the upper portion, where another inscription reads: *Fili qui vere Deus precor his miserere*, "O my son, who are the true God, I implore you to be merciful to this people". The message conveyed by the painting is unambiguous: Mary is represented in such a way as to suggest that, taking advantage of her special relationship to her child, she is exercising her infallible mediation in favour not so much of mankind in general, as one would expect, yet rather of a specific group of individuals, a knight with his wife and daughter. In its aim to suit the

Figure 6.1.2 Cuirassed figure, fragment of a mural painting on the eastern wall of the south transept.

devotional needs of private people, the composition was conceived of as a self-contained, autonomous image.[5]

Scholars basically agree that, at least in an earlier phase, there was no attempt at decorating the columns as parts of a wider figurative cycle.[6] As pointed out by the analysis of stylistic features, the paintings were made by different hands and in different moments in a timespan going from 1130 to the achievement of the mosaic decoration in 1169. A quick look at our scheme (Figure 6.1.6) makes clear that the distribution of images is absolutely

Figure 6.1.3 Virgin Glykophilousa, mural painting, 1130 (see Figure 6, *D5*).

Figure 6.1.4 A knight kneeling at the feet of the Virgin Glykophilousa, mural painting, 1130.

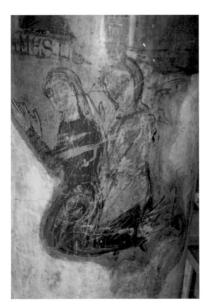

Figure 6.1.5 Two women kneeling at the feet of the Virgin Glykophilousa, mural painting, 1130.

asymmetrical: apparently, the inner row of columns in the north aisle was left completely undecorated, whereas only the first, second, fifth, sixth, seventh, and eleventh columns in the inner south row was embellished with mural icons.

On the contrary, the columns in the central nave were all lavishly decorated. The orientation of the figures is also very peculiar: all those on the south inner row and in the two easternmost columns of the south row of the central nave are displayed on the western half of the shaft (and in two cases, Figures 6.1.6, D5 and D6, also on the eastern one), whereas all others in the central nave are represented facing each other, that is on the south half of the shaft on the northern side and in the opposite way in the southern one.

At least at first sight, the figurative themes seem to have been distributed in a rather incoherent way. They included mostly iconic images, but the prophet Elijah on the eighth column on the southern row of the central nave (Figure 6.1.6, C8) was represented seated in the wilderness and fed by a raven, whereas a *Crucifixion* was displayed on the northern pier close to the entrance to the choir zone. Margaret of Antioch (also known as Marina) was represented twice (Figure 6.1.6, C11, D7), and the Virgin Mary was rendered three times according to different iconographic schemes: apart from the *Glykophilousa* (Figure 6.1.6, D5), the most

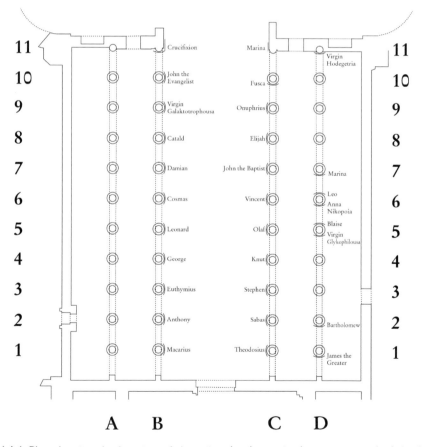

Figure 6.1.6 Plan showing the location of the painted columns in the western end of the Nativity Church (sketch: Thomas Kaffenberger, University of Fribourg).

common *Hodegetria* (holding Christ on her left arm) was represented on the easternmost column of the inner south row (Figure 6.1.6, D11), whereas the much less usual theme of the enthroned Mother of God milking the Child (known in scholarship as *Galaktotrophousa*) was displayed in a no less prominent position on the ninth column of the northern row in the central nave.

As with the *Glykophilousa*, at least another two figures were selected because of the special role they played in their donors' private devotion: James the Greater (Figure 6.1.6, D1) was painted on the initiative of two pilgrims – a man and a woman – who had previously visited the saint's tomb in Santiago de Compostela, as is revealed by the badge associated with the site – a scallop shell – which decorates their bags. Similarly, only somebody from Norway – that is the lady with blonde hair shown in kneeling pose to the left of the image – may be considered to have been responsible for the public display of Saint Olaf's image (Figure 6.1.6, C5), given that the latter's worship was not diffused outside Scandinavia. The same can be assumed for the nearby mural painting representing Saint Knut of Denmark (Figure 6.1.6, C4) and the images of Catald of Taranto (Figure 6.1.6, B8), whose worship had recently developed in the Norman Kingdom of Southern Italy, and Fusca (Figure 6.1.6, C10), whose cult was restricted to the Venetian lagoon. The selection of such local saints bears indirect witness to the involvement of pilgrims from Northern Europe in the embellishment of the nave columns with images manifesting their personal devotion and quest for the sake of the souls. It is not unlikely that also Leonard of Limoges may have represented on the initiative of a private donor, given that this distinctively Western saint, invoked by those who were wrongfully imprisoned, was strongly venerated by pilgrims, who visited his shrine on the way to Santiago and asked for his protection against the risk of being captured during their dangerous voyage to the Holy Land.

It is generally assumed that the first paintings on the columns were made occasionally, as "votive" or *pro anima* images, by will of some devotees who were interested in manifesting their self-dedication to holy intercessors: the latter's representation in a major holy place was considered to reward donors with some outstanding spiritual advantage, since they were meant to be exposed to the sight of many pious people who would have been encouraged to worship those same saints and say a prayer for those who had promoted their public display. On the other hand, it would be misleading to imagine that the decoration started without any involvement of the clergy, and more specifically of the Latin regular canons ruling the basilica. Saints were described, in contemporary liturgical commentaries, as the "pillars" of the church, and their depiction on columns was therefore perfectly acceptable, given that it gave visual shape to this well-known allegorical motif.[7] Moreover, given that the walls of the side-aisles were still covered by precious marble stones belonging to the old Justinianic décors, the large shafts of the columns were the only surface which could be used for mural icons, the upper portions of the walls being reserved for narrative or symbolic themes.

Even if the decoration campaign was originally the outcome of random initiatives, it can be assumed that, in a later phase, efforts were made to include the earliest images into a more coherent program. Not surprisingly, a special emphasis was laid on the embellishment of the two rows delimiting the space of the central nave, which pilgrims were accustomed to go through when entering the basilica. The images displayed in this part of the church were certainly more exposed to the sight of visitors than those in the inner aisles. Furthermore, since the access to the Nativity cave was from the south transept in the Crusader period, the inner south row was certainly regarded as more important than the northern one, which was most probably left undecorated because it was not really involved in the pilgrims' usual itinerary.

Some aspects of the decoration can be better understood if we consider the kinetic approach of ancient beholders to the painted images. Upon entering the nave, they were supposed to simultaneously glance at the murals on both rows of columns, to their left and right. None was intended for an analytical inspection, but their visual interaction came to exert a strong emotional impact on viewers who passed by and were so encouraged to feel that a multitude of holy people was assisting them in their imminent experience of an extraordinary holy site. More than their specific identity, pilgrims could acknowledge, in just one quick look, that they were Orientals and Westerners, local and universal figures, women and men, laymen and clerics, monks and members of the secular church, prophets and relatives of Christ, as well as saints specialised in protecting people against different forms of dangers and calamities.

The great representatives of coenobitic life – Macarius (Figure 6.1.6, B1) and Theodosius (Figure 6.1.6, C1), Anthony (Figure 6.1.6, B2), Sabas (Figure 6.1.6, C2) and Euthymius the Cenobiarch (Figure 6.1.1, B3) – welcomed visitors as soon as they accessed the nave from the narthex and reminded them of the centuries-old tradition of Palestinian and more generally Near Eastern monasticism, which was later to be evoked again, on the southern row, by the images of three figures that were regarded as champions of self-dedication to God as hermits and ascetics, namely Onuphrius (Figure 6.1.6, C9), John the Baptist (Figure 6.1.6, C7) and the prophet Elijah (Figure 6.1.6, C8). The latter was especially relevant in the Bethlehem context, since his most important site of worship in Palestine was the monastery named after him on the way to Jerusalem: given that the complex was deemed to mark the very spot where the prophet had been fed by a raven, it is not surprisingly that this specific episode was evoked in the mural painting. The most important martyrs of the Holy Land, Stephen and George, were displayed not far from the group of ascetics; nevertheless, the display of George in front of Knut on the fourth north column (Figures 6.1.6, B4 and C4) established a clear parallelism between two military saints. Further on, the physical closeness of Leonhard (Figure 6.1.6, B5) to Cosmas and Damian (Figures 6.1.6, B6 and B7) laid emphasis on these three figures' roles as protectors against the risks of life, such as imprisonment and illnesses.

At the end of the north row were the milking Mother of God, John the Evangelist and the only narrative scene, a *Crucifixion* (Figures 6.1.6, B9, B10 and B11). Since it is known that the mosaic program achieved in 1169 included also the latter theme, it can be wondered on which grounds it was also displayed on this precise location, on the pier adjoining the structures of the choir, that is on a liminal position between the nave and the space reserved for the performance of the liturgy. One possible explanation is that its display on this very spot aimed, on the one hand, to evoke the sacrifice of Christ, re-enacted in the Eucharist, in a way that could be easily glanced at by people staying or passing by in the nave, and, on the other, to remind pilgrims of the geographic location of Jerusalem, that is to the north of Bethlehem: as a matter of fact, the northern transept was also later reserved for the representation of the Passion deeds.

On the southern row, the different orientation of the two figures (Fusca and Marina-Margaret: (Figures 6.1.6, C10 and C11) painted on the two last columns was intended to suit the shifting viewpoints of visitors in that area of the nave: it was namely there that pilgrims went from the central nave to the south aisle, whence they could move forward to the transept and the south access to the cave. The decision to display female saints in this area is also perhaps not thoroughly fortuitous, as it may have contributed to suggest the idea of maternity which was expected from the site's experience. Furthermore, Margaret, who was known to have knocked out the devil with her own hands, could work as a powerful reminder of the need to thoroughly purify one's soul before entering that most holy place. The latter could be seen, at one glance, in its closeness to Mary Hodegetria (Figures 6.1.6, D11), who, holding Christ on her right arm, visually synthetised the mystery of Christ's incarnation in a Virgin's womb.

It can be assumed that the original program of decoration extended also to other areas of the church. Possibly, the present-day 19th-century Greek wall paintings displaying saints on the piers joining the choir replaced older mural icons from the Crusader period. There is namely some evidence about the display of kingly figures (possibly Old Testament rulers) on the columns of the altar space.[8] Nevertheless, apart from the official initiatives of decoration, the columns were constantly regarded by pilgrims as an architectural surface where they could leave a sign of their presence. They are namely dotted with graffiti, engraved marks, inscriptions and sketches made by different hands in different periods. Such graphic signs are important not only as historical witness to individual devotion but also inasmuch as they contributed to modify the visual appearance and orient the perception of the columns themselves and the images they displayed.

One of the earliest inscriptions, written in an elegant Arabic script and bearing the date 1192, was included within the image of Saint Fusca, close to her face, in such a way as to visually suggest the supplicant's quest for physical proximity to his special intercessor and wish that his prayer may be heard by her (Figure 6.1.7).[9] The languages employed bear witness to the pilgrims' different origins: if most inscriptions are in Arabic, others are in Armenian, Syriac, Georgian, Latin, Greek, French and German and are only partly published.[10] Western noblemen, in the 15th and 16th centuries, let their coats-of-arms be displayed on

Figure 6.1.7 Arabic inscription close to Saint Fusca's head, 1192 (see Figure 6, *C10*).

the columns of the central row below the saints' mural icons, so that they may be seen and recognised by their fellow citizens upon entering the church. The heraldic shields are often combined with helms, crowns and animal décors, some of them rendered in a very ostentatious way, as with the wolf head on *B1* or the moors' heads and flying ducks on *B7*. Finally, in some cases (as on C5 and C6), holes were carved on the stone surface and arranged in such a way as to shape crosses: most probably, they were used to fix metal votive tablets.

6.2   State of conservation

S. Sarmati

All the columns in the nave, including the corner pillars which connect the nave with the choir, are decorated with paintings on their outer shaft, whereas those in the southern side display paintings on their west and east sides.

In each case, the painted surface corresponds to 2 metres high and 75 centimetres wide rectangular panel. On the panel's perimeter there is a 3 centimetres thick linear frame.

The paintings are icon-like images on stone. A Saint, mostly standing in a frontal position, is represented within each rectangular frame. In three of such rectangles Virgin Mary with Child is represented in three different iconographic types. On the surface of the north wall pillar, which connects the nave with the choir, Christ's crucifixion scene is represented.

Rectangular images are painted on the column's shaft, and the surrounding frame starts from a point being 2,42 metres higher than the column's base and arrives at the column necking.

All the rectangles have the same dimensions, whereas the figures inside have different sizes: some of them are perfectly framed into the column shaft, whereas others are bigger and the upper part of the halo is painted over the top of the column's necking (Figures 6.2.1, 6.2.2).

The background, where the figures are represented, is divided in two halves by a horizontal line, which indicates the transition between the superior part and the floor level; in the upper part, the background is painted with a compact deep-blue colour to represent the sky. On the top of the sky the names of the represented Saints, in both Greek and Latin, were painted on the right and on the left with a brush and a white pigment (Figures 6.2.3, 6.2.4).

The walking surface is represented with a brown transparent colour, a natural earth pigment. Vegetal elements representing grass are painted just in two icons on the floor level (Figure 6.2.5).

The painting's restoration made possible an analytical study of the execution techniques, through the analysis of material evidences in situ and through the laboratory analysis for the identification of materials.

Paintings are created directly on the 4,45-metres-high monolithic columns, made with a compact sedimentary rock, mainly composed of carbonate minerals. The rock's colour, which goes from intense red to yellow ochre, is surely due to the presence of iron oxides (Figure 6.2.6).

In the Middle Ages paintings on stone surfaces were very rare. In those few cases, usually, the technique used is the same as for the masonry support, which consists in applying on the stone a thin layer of plaster or lime, then painted with colours mixed with lime or with the traditional fresco technique (Casaburo, 2017). Paintings on columns of the Basilica of Bethlehem are instead oil paintings, directly made on the stone surface.

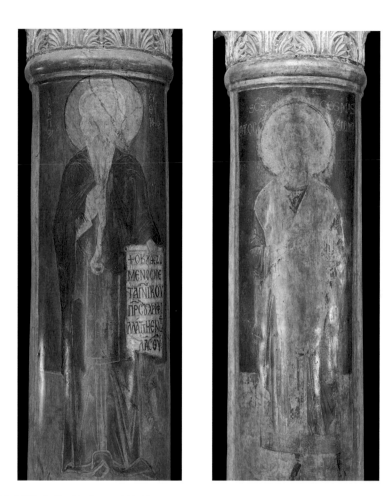

Figure 6.2.1, 6.2.2 St. Euthimius and St. Cosma.

However, even in that period, the technique of painting on a stone surface with an oil binder was already known. In the Heraclius' treatise *De Coloribus et Artibus Romanorum*, the most ancient text on materials and artistic techniques in the medieval period, which survived almost intact, an entire paragraph is dedicated to oil paintings on columns or on stone surfaces.[11]

XXV. **Quomodo preparatur columna ad pingendum**[12]. Si vis aliquam columna vellaminam de petra pingere, in primis optime adsolem vel ad ignem siccare permittes. Deinde album accipies, et cum oleo super marmorem clarissim et eres. Postea illam columnam jam bene sine aliqua fossula planam et politam, de illo albo cum lato pincello superlinies duabus vel tribus vicibus. Postea imprimes cum manu vel brussa de albo spisso,et ita dimittes paululum. Cum vero modicum siccatum fuerit, cum manum tua album planando fortiter retrahes. Hoc tandiu facies donec planum sit quasi vitrum. Tunc vero poteria desuper de omnibus colori bus cum oleo distemperatis pingere. Si vero marbrire volueris, super unum colorem,

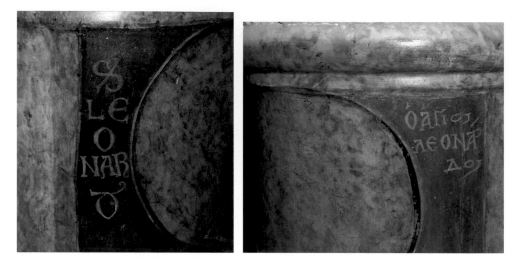

Figure 6.2.3, 6.2.4 St. Leonard both in Latin and Greek.

Figure 6.2.5 Floor level on the painting of Prophet Elias.

vel brunum, vel nigrum, vel alium colorem, cum siccata fuerit marbrire poteris. Postea vernicia ad solem (Merrifield, 1967).

The technique explained in the text is the same as the one assumedly used for the paintings on the Bethlchcm columns. The white hinted at is certainly the white lead which, as proved by the analysis, was largely used upon all the columns. According to Heraclius' instructions, the columns' surface was treated with boiled linseed oil and white lead. In fact, the X-ray fluorescence analysis detected the presence of both materials.

Figure 6.2.6 Sone column, detail.

Because of the bad state of conservation, it is hard to establish the involvement of one or more artists, or of a team, who worked closely together. Anyway, it is clear that the painting technique and the materials used are the same for all the paintings, according to a consolidated compositional scheme of iconographic themes and materials used. The bad state of conservation and the subsequent loss of a large amount of original painting film allowed the restorers to see the underlying preparatory sketch. The different figures were sketched directly on the stone surface with a brush, using transparent colours with dark pigments, such as black and brownish earth colour, without using a *cartone* (cardboard) or other techniques to reproduce images (Mora, L. Mora, P. & Pihilippot , P. 1984).

In the face of Saint Sabas the preparatory drawing of the face's features is clearly visible. Eyes are sketched with yellow ochre, and over and beneath it is possible to see subtle parallel lines, which were used as a guide to place irises properly (Figure 6.2.7).

The preparatory sketch of Saint George's face was also left, and it is full of details. In this case the artists used a dark colour, an earth pigment, applied with liquid transparent brushstrokes, which allowed him to see the young face of the Saint and his thick curly hairs (Figure 6.2.8).

By non-invasive analysis, such as X-ray fluorescence with portable equipment, it was possible to identify the colour palette that was used, mostly composed by an ultramarine blue, largely used on all the icons to paint backgrounds and clothes. Another colour was the white lead which, as already said, was detected upon all the columns. It was used to prepare the stone surface before the execution of the painting or to lighten dark shades, mixed with other pigments.

Figure 6.2.7 St. Saba, preparatory drawing.

In the Virgin Galaktotrophousa painting the typical alteration of the white lead, such as the creation of lead oxide and the blackening of paint film, was also detected (Figure 6.2.9).

Thanks to X-ray fluorescence it was also possible to identify another kind of white colour, a calcium-based white. This pigment was used mostly to replace the white lead in the haloes, mixed with orpiment, which would become otherwise darker if used with lead-based colours or copper-based colours.

The orpiment was used alone or mixed with white to paint the halos, which look golden thanks to the brightness of the orpiment.

The yellow ochre was largely used to paint the clothes or mixed with white to paint the skin tone in all the paintings (Figure 6.2.10).

Brown soils, such as raw umber and green umber, were largely used for dark clothes.

In the image of the Crucifixion, cinnabar ore was detected, on the perimeter of the figures. Except for this case, where cinnabar ore was found, red colours are always iron oxides. The use of cinnabar ore was unusual and it could testify to the presence, in that case, of a different artist.

Fragments of a gold leaf were found on the crown of St. Olaf. The leaf had been applied on the stone surface with a resin (Figure 6.2.11).

Figure 6.2.8 St. George, preparatory drawing.

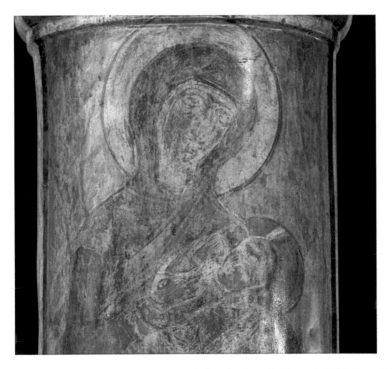

Figure 6.2.9 Virgin Galaktotrophousa, alteration of white lead on the face of the Virgin and the child.

Figure 6.2.10 St. Saba, yellow ochre.

Colours were used pure or mixed with white for homogenous compact highly pigmented backgrounds. Then they worked on these backgrounds by overlapping different layers to create clothes' folds, face features or shadows. In the end light veils of transparent colour were applied (Figure 6.2.12).

Paintings are very damaged due to the fragility of the execution technique: oil paintings on stone surfaces are, in general, fragile artefacts.

Because of the oils' polymerisation and the environmental thermo-hygrometric stress, the painting film can become stiff, break, loose adherence to the stone surface and fall down.

Moreover, when the basilica was conquered by Ottomans, paintings were vandalised: saints' faces were erased by scraping out the colour from the columns, and just some of the faces of the Virgin were spared.

Figure 6.2.11 St. Olaf, golden fragment on the crown.

Figure 6.2.12 St. James the Greater, detail.

Figure 6.2.13 St. Theodosius.

There are no reports of previous restorations of painted surfaces. However, until a few years ago a regular periodic maintenance was carried out in order to brighten up icons' colours on columns. During the maintenance, petroleum and wax mixtures were applied on the columns' painted surface, obfuscated by candles' smoke and dust.

Two are the last reported restoration. The first is the one carried out by the Department of Antiquities of the State of Israel in 1946. There are no documentations at this regard. What is known is that a cleaning process of the paintings was executed in order to make them more legible. The most recent is the one carried out by Gustav Künhel during research for his PhD thesis in 1976. He cleaned and photographed the paintings to study them, and the results were illustrated to Tel Aviv University as his PhD dissertation.[13]

Figure 6.2.14 St. George.

Nevertheless, in the beginning of the present restoration work, paintings were again unreadable because they were covered by a thick layer of dirt caused by the oxidation of old varnish and by deposits of candles smoke, which prevented the view of the original colours.

Saints' faces and lots of other decorative details are irretrievably lost. Faces' features can be seen only from the preparatory sketch and from painting film fragments which are still present on the stone (Figure 6.2.13).

Clothes that remained lost most of their surface finishes, which enriched their patterns. Some icons completely lost their painting film, but the figure can be still identified, thanks to the outline of the shape left on the columns (Figure 6.2.14).

In some areas painting film was lifted and rippled. However, despite the impossibility to see the original colours and despite the great loss of original material, paintings showed good adhesion to the substrate.

References

Casaburo, M. (2017) *Pittura su pietra*. Nardini editore: Firenze.
Merrifield, M. (1967) *Original Treatises on the Arts of Painting*, Volume 1, cap. III. The Getty Conservation Institute, New York, pp. L–LII.
Mora, P., Mora, L. & Philippot, P. (1984) *Conservation of Wall Paintings*. Butterworths: London.
Wilkinson, J. (1977) *Jerusalem Pilgrims before the Crusaders*. AbeBooks: Jerusalem.

6.3   Interventions

S. Sarmati

The painting restoration was a conservative work, carried out according to the principle of the minimum intervention, in order to maintain the traces of the past but, at the same time, stop the degradation process.

Some preparatory activities, such as the survey of the painting materials, close observation of the surfaces both with direct and grazing light, and subsequent mapping of the most critical areas affected by painted film lifting or loss were carried out before the restoration work.

Colour fragments flaking off the stone surface were reattached by syringe injections of a micro-emulsion of acrylic resins.

The cleaning method was chosen, taking into account the specificity of the state of conservation of the paintings.

The smoke deposits were removed without completely removing the overlying varnishes. The varnishes were just lightened, in order to appreciate again the original colours of the paintings and, at the same time, to leave on them the coating created over the centuries (Figure 6.3.1).

The cleaning process was carried out gradually, using high-viscosity solutions to remove the superficial dirt and, at the same time, to prevent a deep penetration as well as its evaporation of the solvent.

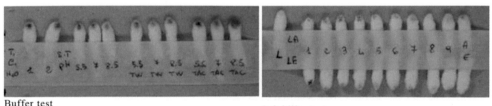

Buffer test
T.C. H2O = water compatibility test
S.T.= Buffer solution pH 5.5, 7.0, 8.5.
TW= buffer solution with TWEEN 20
TAC= buffer solution with tribasic ammonium citrate

Solubility test
L A = Ligroin + Acetone
L E = Ligroin + Ethanol

Figure 6.3.1 Buffer and solubility test.

Figure 6.3.2 Buffer solution test.

Tests with buffer solutions were carried out in order to verify the material's resistance in an aqueous environment, and a solubility test was performed in order to identify the most effective solvent mixture for dirt removal (Figure 6.3.2).

The presence of a thick superficial deposit was detected.

Tests showed that the deposit could be removed completely with a buffer solution pH 8.5 with a weak chelating agent. Instead, the solvent mixture of ligroin and acetone with fd 91 only partially removed the superficial deposit.

Taking into account the tests results, it was decided to use a weakly basic (pH 8.5) buffer solution of a chelating agent in water/oil emulsion (W/O), effective to remove dirt without damaging the painting surface.

Two emulsions were prepared:
W/O: W = Buffer solution pH 8.5 + Triammonium Citrate
O = Ligroine (surfactant used: Brij 35 + Tween 20);
Velvesil Plus + 10% Buffer Solution pH 8.5 + Triammonium Citrate

The cleaning test was carried out on the painting representing prophet Elijah, more precisely on the deep-blue background, near to the right side area.

Firstly, the cleaning process was performed on a little area with the first emulsion. It was applied with a soft brush, and the application time was about 3 minutes; the surfactant

Figure 6.3.3 The application of emulsion on surface.

residuals were removed with ligroin by using a cotton swab, and the whole surface was treated with a cotton swab slightly moistened with demineralised water (Figure 6.3.3).

A second cleaning was carried out by applying a pH 8.5 buffered solution of the weak chelating TAC (10% in weight) with Velvesil Plus silicone gel. The quantity of products used does not give rise to an authentic emulsion, but it enables a light superficial cleaning. The contact time lasted about one hour. The gel residuals were removed with pure Ligroine followed by Cyclomethicone D5 (Figure 6.3.4).

In the areas where most of the painting film was lost, the few fragments of colour showing a good adhesion to the surface were dry cleaned with a PVC-free rubber without phthalates.

The heavily damaged old grouts inside little holes and cracks were removed (Figure 6.3.5).

The gaps were filled with hydraulic lime Saint-Astier mortar and with various limestone powders, trying to create a mixture whose colour was as close as possible to the stone surface colour.

The painting reintegration, made with watercolours, simply consisted of lightening the colour's shade of the abrasions and of the inhomogeneous areas of the painting surface with veils of colour (Figure 6.3.6).

On the surface was applied the varnish Regalrez, a low molecular weight aliphatic polymer with a low oxidising power.[14] The varnish was applied with a brush on the reintegrated surfaces as a final protective.

Figure 6.3.4 Cleaning test.

Figure 6.3.5 Mechanical removal of old grout.

Figure 6.3.6 The painting reintegration:

Notes

1 M. Bacci, *The Mystic Cave: A History of the Nativity Church in Bethlehem*, Brno, Rome, 2017, pp. 125–126.
2 Bacci, *The Mystic Cave*, pp. 91–92 and 96–102.
3 Eutychius of Alexandria, *Annals*, M. Breydy (ed.), *Das Annalenwerk des Eutychius von Alexandrien*, Leuven, 1985, p. 120.
4 Daniel the Higoumen, K.D. Seemann (ed.), *Khozdhenie – Wallfahrtsbericht. Igumen Daniil. Nachdruck der Ausgabe von Venetinov 1883/85 mit einer Einleitung und bibliographischem Hinweisen von Klaus Dieter Seemann*, Munich, 1970, p. 64.
5 G. Kühnel, *Wall Painting in the Latin Kingdom of Jerusalem*, Berlin, 1988, pp. 15–22; J. Folda, *The Art of the Crusaders in the Holy Land 1098–1187*, Cambridge, 1995, pp. 91–96; J. Folda, *Crusader Art: The Art of the Crusaders in the Holy Land, 1099–1291*, Aldershot, 2008, p. 28; J. Folda, Twelfth-century Pilgrimage art in Bethlehem and Jerusalem: Points of contact between Europe and the Crusader Kingdom. In: Bacile, R.M. & McNeill, J. (eds.) *Romanesque and the Mediterranean: Points of Contact Across the Latin, Greek and Islamic Worlds c. 1000 to c. 1250*, Leeds, 2015, pp. 2–4; Bacci, *The Mystic Cave*, pp. 131–132.
6 R.P. Germer-Durand, Décoration de la basilique de Bethléem. *Échos de Notre-Dame de France*, (1891), 41–45; H. Vincent & F.-M. Abel, *Bethléem. Le sanctuaire de la Nativité*, Paris, 1914, pp. 168–176; L. Dressaire, Les peintures executées au XIIe siècle sur les colonnes de la basilique de Bethléem. *Jerusalem*, 27 (1932), 365–369; R.W. Hamilton, *The Church of the Nativity, Bethlehem: A Guide*, Jerusalem, 1947, pp. 69–81; V. Juhasz, Las pinturas de los Cruzados en la basilica de Belén. *Tierra Santa*, 25 (1950), 313–318 and 349–353; V. Juhasz, *Pinturas y grafitos de la Basílica de la Natividad en Belén*, forthcoming; B. Bagatti, *Gli antichi edifici sacri di Betlemme in seguito agli scavi e restauri praticati dalla Custodia di Terra Santa (1948–1951)*, Jerusalem, 1952, pp. 93–106; J. Folda, Painting and sculpture in the Latin Kingdom of Jerusalem 1099–1291. In:

Hazard, H.W. (ed.) *The Art and Architecture of the Crusader States*, Madison, WI, 1977, pp. 251–280 and 121–122; G. Kühnel, Das Ausschmückungsprogramm der Geburtsbasilika in Bethlehem. Byzanz und Abendland im Königreich Jerusalem. *Boreas*, 10 (1987), 133–149 and 143–146; Kühnel, *Wall Paintings*, pp. 5–147; Folda, *The Art of the Crusaders*, pp. 91–97, 163–166, 283–284, 315–318, 364–371, 462; G. Kühnel, Crusader monumental painting and Mosaic. In: *Knights of the Holy Land: The Crusader Kingdom of Jerusalem*, exhibition catalogue (Jerusalem, The Israel Museum, Summer–Fall 1999), Jerusalem, 1999, pp. 202–215 (here 208–210); Folda, *Crusader Art*, pp. 28, 54; Folda, Twelfth-century Pilgrimage art; Bacci, *The Mystic Cave*, pp. 132–136.

7 See especially R. Suntrup, Das Kirchengebäude in der allegorischen Liturgieerklärung des Durandus von Mende. In: Wagner, D. and Wimmer, H. (eds.) *Heilige: Bücher-Leiber-Orte. Festschrift für Bruno Reudenbach*, Berlin, 2018, pp. 299–309.

8 Bacci, *The Mystic Cave*, pp. 143–144.

9 Kühnel, *Wall Paintings*, p. 102; Folda, *The Art of the Crusaders*, p. 473.

10 The graffiti are listed, but only partly transcribed in Stone 1992–1994, also online at http://rockinscriptions.huji.ac.il/site/index (retrieved 27 November 2018). A systematic publication of Western inscriptions and graffiti in Bethlehem is provided by Juhasz, *Pinturas y grafitos*; cf. also Vincent and Abel, *Bethléem*, p. 189 note 1; D. Kraack, *Monumentale Zeugnisse der spätmittelalterlichen Adelsreise. Inschriften und Graffiti des 14.-16. Jahrhunderts*, Göttingen, 1997, pp. 135–149; G. Gagoshidze, Georgian inscriptions in the holy land. In: *Georgians in the Holy Land: The Rediscovery of a Long-Lost Christian Legacy*, London, 2014, pp. 67–91 (here 71–74), publishes the graffiti in Georgian script, whereas there a project concerning the Armenian graffiti is presently ongoing (*Armenian Inscriptions of the Land of Israel and the Sinai*, coordinated by Michael Stone). A systematic inquiry of Arabic inscriptions is still a desideratum.

11 M.P. Merrifield, *Medieval and Renaissance Treatises on the Arts of Painting*, original texts with English translations, Courier Corporation, 1999.

12 *XXV. If you wish to paint on a column or slab of stone, first let it dry very perfectly in the sun or before a fire. Then take white, and grind it very finely with oil upon a marble slab. Afterwards, the column being well smoothed and polished, without any crevices, lay on in two or three coats of that white, with a broad paintbrush. Then rub very stiff white over it with your hand or with a brush, and let it remain a short time. When tolerably dry, press your hand strongly over the white surface, drawing your hand towards you. Continue to do this until it with your as smooth as glass. You will then be able to paint upon it with all colours mixed with oil. But if you wish to imitate the veins of marble on a general tint (brown, black, or any other colour) you can give the appearance, when the ground so prepared is dry. Afterwards varnish in the sun.*

13 A. Keshman, Crusader Wall Mosaics in the Holy Land. Gustav Kühnel's work in the Church of the Nativity in Bethlehem. *Arte Medievale*, S.IV, 3 (2013), 249–262.

14 F. Talarico, Proprietàdelleverniciidrocarburiche e prospettive. *Kermes*, 72 (Ottobre–Dicembre 2008), 70–73.

References

Cremonesi, P. (2012) *L'ambiente acquoso per il trattamento di opere policrome*. Il Prato Padova.

Dorge, V. (2004) *Solvent Gels for the Cleaning of Works of Art: The Residue Question*. Getty Conservation Institute, Los Angeles.

Wolbers, R. (1992) Recent development in the use of gel formulation for the cleaning of paintings. In: *Restoration 1992 Conference Preprints*. UKIC74-75, London.

Wolbers, R. (2000) *Cleaning Painted Surfaces: Aqueous Methods*. Archetype Publications, London.

Chapter 7

Exterior wall surfaces

7.1 Materials and their state of conservation

S. Sarmati and G. A. Fichera

7.1.1 Archaeological analysis of the architectures

The archaeological analysis of the architectures of the basilica (see Section 3.7) made it possible to define with certainty the constructive sequences of the monumental complex through the definition of all the constructive and destructive actions that have marked life over the centuries. In order to achieve this objective, in addition to focusing on the main relationships of relative chronology between the individual masonry walls, was also of fundamental importance an extremely analytical sampling of the construction techniques used.

The analysis of the masonry walls and of the construction techniques adopted was performed almost exclusively on the external wall of the basilica, as the interiors are covered with mosaics or plaster that hide the masonry work. A large 2-metre-high fragment of masonry in the southern side aisle had no plaster covering and showed technical characteristics similar to the external ones. This masonry was also covered with some graffiti of difficult chronological assignment but certainly not dating back to the original construction phase of the basilica.

For the identification and classification of masonry techniques, the parameters adopted were primarily the construction material, the type of preparation and the refinement of the individual elements, installation and their size, as well as the type of binder used.

The harmony of the construction process, as well as the relationships between the wall structures, which show the substantial contemporaneity of the entire construction, is further confirmed by the adoption of construction techniques characterised by strong homogeneity and high technological level. These considerations reveal the employment of highly professional workers, as it should be for an imperial assignment.

The construction builders of the church resorted almost exclusively to a single technique, which has very few variable elements. Indeed, the walls are characterised by perfectly squared ashlars of large dimensions (up to 1 m length and 0.4 m height) placed on horizontal and parallel rows of fairly regular heights depending on the heights of the segments[1] (Figure 7.1.1).

The traces of the working tools indicate flat-blade chisels, with a thickness of 2/3 cm, as the instruments adopted for the contour realisation of the concave (anatirosis) and two

a

Figure 7.1.1 Construction technique sample.

different types of "gradina" (tooth-blade chisels) employed for the surface polishing of the segments (Figure 7.1.2 b–c).

The traceable joints and laying beds are extremely thin, thanks to the perfectly fitting faces of the ashlars. The placement of cement-based mortars, necessary to cover the original laying beds, has not completely obliterated the traces of the original ones, which appear to be very compact whitish lime mortars rich in shredded brick fragments. Such fragments, which even cover the edges of the individual blocks, were smoothened to fill the small natural fractures of the stone and were sometimes engraved with a pointed tool (Figure 7.1.2d–e).

The technique described above is always associated with the local lithotype of malaki stone, with nuances of light yellow to grey and with different levels of compactness, and it can be ascribed to the industry of highly qualified workers at the service of the imperial client.

The contemporary adoption of this technique is supported by precise stratigraphic relationships, proving that all the walls of the basilica were built within a homogeneous project, with very special expedients, such as the construction of L-shaped ashlars in order to link perpendicular walls together (Figure 7.1.3a).

Figure 7.1.2 b-c. Traces of working tools; d-e mortar joints with brick fragments.

Figure 7.1.3a Angle between the walls of the central nave and the north transept. The semi-pillar with segments specially worked in the form of "L" or with more elaborate forms connect perpendicular walls to each other.

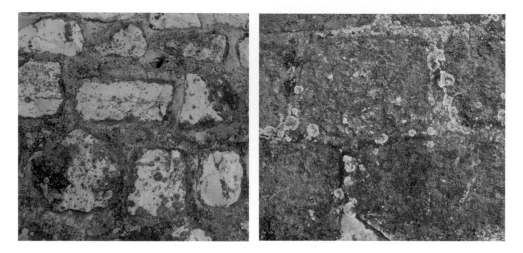

Figure 7.1.3b, 7.1.4 Biological deposits.

7.1.2   State of conservation of the surface

The state of conservation of the surface was severely compromised due to the de-cohesion of the stones and the presence of widespread alveolation of the stone surface, especially in the areas where wind power is greater.

Biological deposits consisting of algae and lichens were present almost everywhere, with thick patinas, especially in north-facing areas (ICOMOS 2011).

The biological growths on the stone were particularly extensive and compact, with algae and lichens of various types and colours. These organisms strongly contribute to the deterioration of the stone, altering its original colour.

Biological patinas on the stone are well-known phenomena and have been long studied as one of the most relevant mechanisms of stone degradation (Figure 7.1.3b, 7.1.4).

In some areas that are protected by the washing action of rain, compact deposits of black crusts were present. Air pollution is the main culprit in the decay of stones. Acid rain is responsible for deposition of acid pollutants on limestone with the formation of black crusts. If the stone is exposed to regular rain flows, the reaction products are washed away, and the surface of the stone gradually deteriorates by pulverising. However, if the stone is in a relatively sheltered position, the reaction products accumulate and can form dense black crusts on stone surfaces (Figure 7.1.5, 7.1.6).

The soluble salts represent an additional cause for stone decay. Salts can damage the stone in several ways. The most relevant is salt crystal growth within the pores of the stone, which generates enough stress to overcome the stone tensile strength, turning it into dust. The deterioration caused by soluble salts mostly occurs in external environments, where the stone is subject to rain cycles with subsequent drying.

The entire stone surface was covered with a dark ochre patina due to the oxidation of protective dusters applied in previous interventions.

The most relevant damage to the stone surface was the extensive stucco covering with cement and other materials carried out during previous maintenance operations. In some areas these fillings completely covered the stone surface, hiding the masonry (Figure 7.1.7).

Figure 7.1.5, 7.1.6 Black crust.

Figure 7.1.7 The fillings completely covered the stone surface.

References

Bacci, M., Bianchi, G., Campana, S. & Fichera, G. (2012) Historical and archaeological analysis of the Church of the Nativity, in the Church of the Nativity in Bethlehem: An interdisciplinary approach to a knowledge-based restoration. *Journal of Cultural Heritage*, 13, 5–26.

Bianchi, G., Fichera, G. & Campana, S. (2016) Archeologia dell'Architettura nella Basilica della Natività a Betlemme, pp. 1567–1589, in Olof Brandt – Vincenzo Fiocchi Nicolai (a cura di), Acta XVI Congressus Internationalis Archaelogiae Christianae Romae (22–28.9.2013). *Costantino e i Costantinidi. L'innovazione costantiniana, le sue radici e i suoi sviluppi*.

ICOMOS-*ISCS* (2011) *Illustrated Glossary on Stone Deterioration Patterns.* Manual. ICOMOS, Paris.

7.2   Intervention techniques on stones and mortar joints

S. Sarmati

The interventions on the stone surfaces started immediately after the roof renovation and involved the restoration of the external walls of the basilica.

All interventions on the stone have been preceded by intense investigations, since the restoration process is about knowledge above all. A laser scanner record of all the internal and external surfaces has been created in order to have an adequate graphic basis for the documentation of both the initial state and the subsequent restoration interventions. Investigations on the type of stone and on the mortars put the basis for a systematic study of the walls' structure, allowing the recognition of the various historical periods.

The first treatment performed was the application of biocides product on all surfaces. The stone was sprayed with a concentrated liquid preparation of quaternary ammonium salts with a broad spectrum of activity, covering fungi, bacteria and algae (Figure 7.2.1).

Subsequently careful cleaning of the surfaces was performed, eliminating all the necrotised deposits and black crusts where present. Prior to the cleaning procedure, the entire

Figure 7.2.1 The application of the biocide.

stone surface was covered with a dark ochre patina due to the oxidation of protective dusters applied in previous restoration efforts. The cleaning carried out with ammonium carbonate has removed the black crusts and the deposits, maintaining the light ochre patina of the oxalates on the stone (Figure 7.2.2).

The stone surface has been consolidated by imbibitions with ethyl ester of silicic acid (ethyl silicate) applied especially in the highly degraded areas, where pulverisation occurred. Thanks to its formulation the ethyl silicate has excellent penetrating power, able to reach the full stone depth (Figure 7.2.3).

Figure 7.2.2 The oxalates patina.

Figure 7.2.3 The application of the ethyl silicate.

Figure 7.2.4 Removal of cement fillings.

Figure 7.2.5 Four filling tests.

Figure 7.2.6 The south façade after restoration.

All the incoherent and non-pertinent fillings performed with inappropriate materials have been removed and properly replaced. Several tests have been pursued in order to select the most convenient materials adapted to the preservation of the colour and the morphology of the mortars, also based on the mineralogical analyses performed on the original samples (Figure 7.2.4).

The new inlays were performed with hydraulic limes, employing local stone dust and brick dust. Several tests have been necessary to reproduce the aspect of the original stone (Figure 7.2.5).

The restoration has reinstated the original wall surface condition without eliminating former restoration attempts performed over the centuries (Figure 7.2.6).

Note

1 (Fichera in Bacci *et al.*, 2012; Fichera in Bianchi *et al.*, 2016).

Chapter 8

Wooden architraves

C. Alessandri, N. Macchioni, M. Mannucci,
M. Martinelli and B. Pizzo

8.1   Internal geometry, material and technical characteristics

Wooden architraves are located above the four colonnades of the nave and the aisles and above the columns of the north and south corners (Figures 8.1.1, 8.1.2).

Prior to the present-day restorations, such architraves were mostly concealed by plaster or, in some cases, by wooden boards added in recent times. These architraves, made of Lebanon cedar, were introduced as tie beams whose basic purpose was to connect in some way the columns together, to absorb the horizontal thrust components at both ends of each colonnade and possible horizontal forces produced by earthquakes. Moreover, in the intentions of the designers, they had to guarantee the availability of the widest possible surfaces, not interrupted by arches, on which the mosaic decorations could be carried out. Besides their structural role, wooden architraves have also an important aesthetic value because of the decorations and carvings present on them, which appear to be original. The central element is decorated at the intrados, whereas the two lateral ones are decorated on their external sides (Figures 8.1.3, 8.1.4).

Figure 8.1.1 Architraves in the nave and aisles.

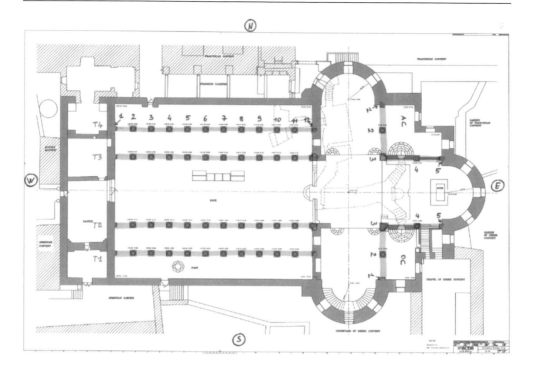

Figure 8.1.2 Architraves (highlighted in purple) above the colonnades.

Figure 8.1.3 Decorations at the intrados.

As evidenced by inspections, these wooden architraves are constituted by three Lebanon cedar wood beams set side by side and usually spanning over three consecutive columns. They are surmounted by four smaller elements, two of which make up the external cornices (Figure 8.1.5). In some cases, the central beam is composed by two superimposed elements, the lower of which has a height about one half of that of the upper element (Figure 8.1.6).

Figure 8.1.4 Side decorations.

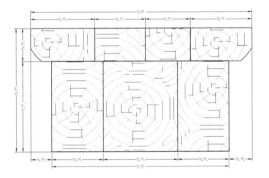

Figure 8.1.5 Usual architrave cross section.

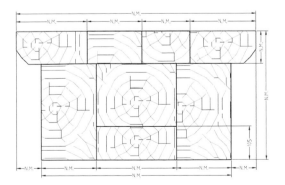

Figure 8.1.6 Particular architrave cross section.

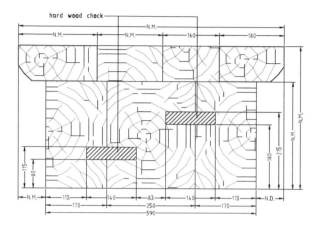

Figure 8.1.7 Horizontal chocks.

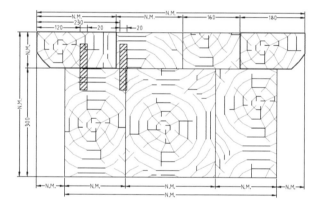

Figure 8.1.8 Vertical chocks.

All these elements are also connected to each other by means of nails and internal wood chocks, made of hardwood (probably oak) and inserted both horizontally and vertically in order to prevent relative movements (Figures 8.1.7, 8.1.8). No connectors are visible from the outside.

As said previously, although they may appear continuous at first glance, the architraves, or at least their external beams, are always head joined on the third, sixth and ninth columns by half dovetail joints (Figures 8.1.9, 8.1.10).

On the contrary, the results of the instrumental analyses carried out show that the central beams are probably interrupted also on different columns. However, they are also head-joined by means of half dovetail joints or probably by other kinds of joints if they are composite beams (Figure 8.1.11).

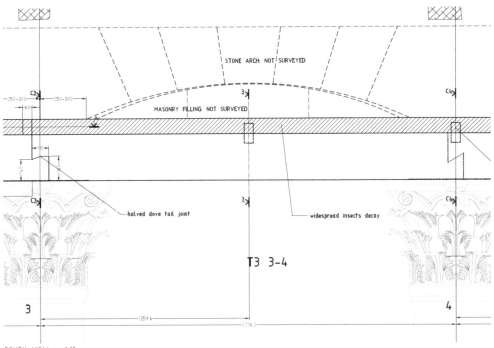

Figure 8.1.9 Half dovetail joint between external beams and internal chocks.

Figure 8.1.10 External half dovetail joint.

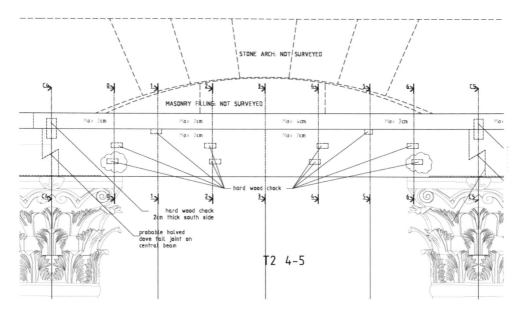

Figure 8.1.11 Half dovetail head joints between internal beams. Internal chocks are also visible.

The angle joints in the corners are slightly more complex as each peace is differently connected to the other and the actual connections are not visible from the outside except for the external elements. Therefore, only hypotheses can be made about such joints (Figure 8.1.12).

It must be noted that the weight of the walls is not borne by the architraves but by low, relieving arches, partly walled up and hidden under the plaster and springing from the architraves themselves in correspondence of each column (Figure 8.1.13). Thus, inside each span the architrave bears only the weight of the filling material between arch and architrave, usually made up of loose stones (Figure 8.1.14), whereas in the parts above the capitals it is compressed perpendicularly to the wood grain. There is no vertical connection between capital and architrave. The latter simply rests on a circular relief (diameter 60 cm) of the upper base of the capital in axis with the column.

In correspondence of the north and south corners, where each architrave joins the orthogonal one, a reinforcement is provided by external L-shaped steel bars nailed to the beams (Figures 8.1.15 a, b and 8.1.32).

It must be remembered that the dendrochronological analyses carried out in 2010 during the first investigation campaign [1][2] in the first row of architraves of the south aisle showed that they are made of cedar dating back to the beginning of the seventh century with an evaluation error of about +/− 60 years. Therefore, there are sufficient reasons to believe that they are coeval with the construction of the church that occurred in the second half of the sixth century. In this case, they would be, together with the wooden tie beams of Hagia Sophia in Constantinople, among the most antique wooden elements in the Mediterranean area.

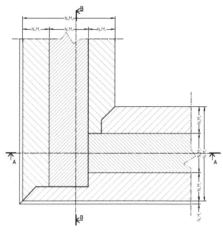
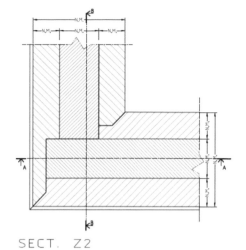
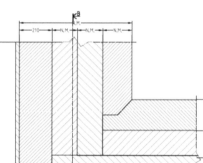
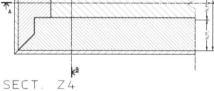
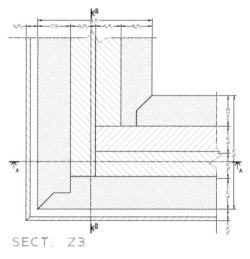

SECT. Z1

SECT. Z2

SECT. Z3

SECT. Z4

Figure 8.1.12 Possible joints at the corners.

Figure 8.1.13 Details of the relieving arches during restoration.

Figure 8.1.14 Filling material in detail.

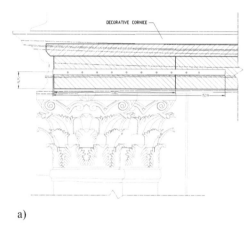

a)

b)

Figure 8.1.15 External L-shaped steel bar – (a) lateral view, (b) front view.

8.2   External decorations

On both sides of the architraves the ornaments include, from top to bottom, a smooth listel, a frieze of rinceaux including roundish fruits, a sequence of globular motifs (Figures 8.1.4, 8.2.16), and two bands of rosettes shaped by leaflets arranged in such a way as to form lozenges (Figures 8.1.4, 8.2.17). The artist relied on a repertory of forms which was rooted in the artistic traditions of the Christian Near East [3]: the lozenge-shaped rosettes compare, for example, with fifth- and sixth-century Coptic stone lintels, whereas the foliate motifs may be understood as definitely stylised variants of the friezes with acanthus leaves housing pomegranates, which are also frequent in Egypt.

At the intrados, the central beam is decorated with a band of interlacing scrolls shaping medallions where acanthus leaves are symmetrically arranged around eight-petal rosettes (Figure 8.2.18). In the middle of this band stands a laurel wreath (or crown of victory) housing a cross (Figure 8.2.19).

Figure 8.2.16 Frieze of rinceaux and globular motifs.

Figure 8.2.17 Double bands of rosettes.

Figure 8.2.18 Central band of interlacing scrolls.

Figure 8.2.19 Laurel wreath.

Some of the architraves display a particularly elegant variant where the scrolls spring out of cornucopiae and are embellished with pearl-dotted borders. The combination of such motifs is frequently encountered in both Egypt and the wider area of Syria and Palestine and was appropriated also in the decoration of early Islamic monuments in Jerusalem in the Umayyad period [3].

8.3   State of conservation and past interventions

The state of decay was various and diversified: some parts were missing, there were cracks, incoherent deposits, material deterioration due to fungi and insect attacks, spots due to water percolation. The state of decay worsened also in consequence of unsuitable interventions made in the past, like painting of capitals and wooden surfaces, some inappropriate superficial treatments and also in consequence of human negligence that produced visible anthropic damages.

In general, the mechanical quality of wood was not worrisome, except for some (few) cases where beams showed clearly visible defects (knots, fiber deviation, etc.) and extended material decay. Decorative (carved) surfaces on lateral (vertical) sides of the architraves were also (visually) evaluated. The lower side was not considered because it generally resulted in quite good conditions, whereas the lateral sides showed a wide range of physical conditions, from very good conditions (0% degradation, that means with intact carvings or with negligible damages) to very bad conditions (100% degradation that means carvings completely destroyed because of decay, anthropic interventions etc.).

In 2011 the Consortium had given some guidelines for the restoration of those architraves where the most severe structural damages had occurred [1].

On November 26, 2015, the contractor's spacialized team started with additional surveys in order to update all the data concerning the state of conservation of the architraves. X-ray scanning and resistograph inspections, visual analyses and sound tests were carried out. These analyses confirmed that lateral sides of some architraves were seriously damaged, and in the past they had been covered by plaster or wooden boards in order to hide the damaged

Figure 8.3.20 Cast of the side decorations.

parts. Before starting the diagnosis both plaster and boards were removed. In some cases, it was possible to have perfect casts of the side decorations (Figure 8.3.20). These casts were preserved together with other finds.

All on-site inspections were carried out without removing original wood parts if not extremely decayed. The results of the diagnosis were used to define the best intervention techniques. The inspection was extended to the entire architrave system (nave, aisles, transepts/ apse and corners), with special focus on the geometrical details of connections (both visible and not visible), inner and outer decay, wooden species and mechanical quality of wood.

Beams within each span of nave and aisles were identified by the number of the architrave (colonnade) and by Arabic numbers from 1 to 11 starting from the west side (façade). For instance, T2-03 identifies the south architrave of the nave (see Figure 8.1.2), span between the third and the fourth columns. Beams of the corners were numbered from 1 to 4 anticlockwise in the Armenian corner and clockwise in the Orthodox corner (Figure 8.1.2). As each beam is composed of different elements, each element is identified by its position referred to the four cardinal points and to the central axis of the composite beam.

The number and position of the performed "resistographic" drillings were decided according to a predetermined scheme; the drillings were usually performed at the bearing cross sections, at the mid-span cross section and at the quarters of the span if necessary; a variable number of drillings were performed on each cross section perpendicularly to the beam axis. With reference to this basic scheme, the number and distribution of drillings were increased on the basis of visual inspections, hammer responses and drilling findings as well, according to a feedback procedure.

All inspections were performed from 27 November to 16 December 2015 according to Italian Standard UNI 11119:2004 [4]. Drillings were carried out by means of an IML Resi

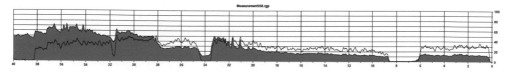

Figure 8.3.21 Drilling test profile.

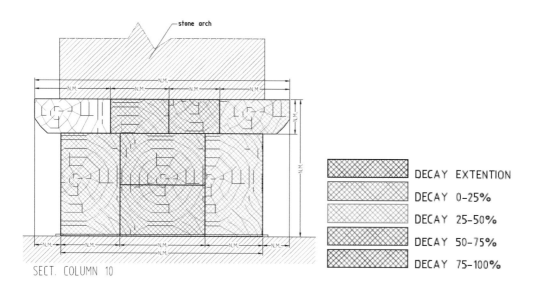

Figure 8.3.22 Example of the different levels of damage for column 10 of the south aisle.

B400 drilling device + electronic unit (for detailed description see www.iml.de). The instrument used was electronically controlled and capable to reach 40 cm depth with a resolution of 0.02 mm. Outputs of "resistographic" drilling tests are available as x-y diagrams named "profiles" (Figure 8.3.21) where the x parameter is the "drilling depth" in cm and the y parameter is the "drilling resistance" of wood (actually the electrical power absorption).

A total of 1372 drilling tests were performed.

In each lintel the areas affected by material decay were represented both in the longitudinal direction and in transverse sections. In the latter case, as already done in the roof trusses, the state of decay was represented by means of four ranges of decay., i.e. 0–25%, 25%–50%, 50%–75%, 75%–100% (Figure 8.3.22). Data are reported at steps of one-fourth span.

The state of decay of beams and cornices was also represented, with the same ranges of percentage decay, in two different plans, one referred to the cornices and the other to the beams.

Several alterations were observed due to biological decay (Figure 8.3.23) and anthropic interventions (Figure 8.3.24); the main problems dealt with fungi (rot), insects and water percolation.

Decay of beam elements varied from negligible to severe, sometimes in the same composite beam, as in Figure 8.3.22. In some cases, cornice elements and side surfaces, completely destroyed, in recent times had been covered by plaster, as in the nave (Figure 8.3.25), or

Figure 8.3.23 Biological decay.

Figure 8.3.24 Anthropic damage.

replaced by wooden boards (Figure 8.3.26) after cornice and decoration had been canecelled (Figure 8.3.27a). Elsewhere the traces of the material decay had simply been left in sight, visible signs of the passing of time and human negligence.

Some beams had partially collapsed due to a severe decay of their components or showed a visible crushing of the part above the capitals due to the loads transferred by the arches (Figure 8.3.27b). Consequently, the superimposed structures (walls and stone arches) showed significant failures (Figure 8.3.28).

In some cases, internal decayed elements had been replaced by new elements (generally of a different species, frequently oak); stone elements had been inserted in place of wood above some column capitals, so interrupting the material and structural continuity of the

Figure 8.3.25 Architraves in the nave.

Figure 8.3.26 Added wooden boards and cornices.

Figure 8.3.27a Cancelled cornice and decorations.

Figure 8.3.27b Crushing of the architrave.

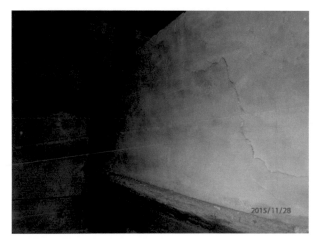

Figure 8.3.28 Fracture in the masonry wall.

Figure 8.3.29 Stone element above the capital.

Figure 8.3.30 Stone element above the capital.

architrave (Figures 8.3.29, 8.3.30); iron bars had been laterally applied to join beam elements to each other (Figure 8.3.31) or to reinforce the corner joints (Figure 8.3.32).

The most important problems in beams and cornices were located between the 11th and the 12th columns in both the north and the south aisles (Figures 8.3.33, 8.3.34). In both cases, the decay extended starting from the walls in which the two 12th columns were included, suggesting that the problem originated from water percolation along these walls. In these cases, the decay consisted of a physical lacuna involving part of all the three architrave elements (the central and the two lateral ones). Whenever possible, the remaining decorated parts were all preserved.

Figure 8.3.31 Nailed iron bar.

Figure 8.3.32 Nailed iron bar.

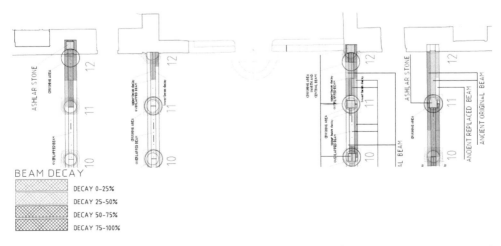

Figure 8.3.33 Examples of decay in lintel beams, starting from the 12th column and extending towards the 11th and even the 10th.

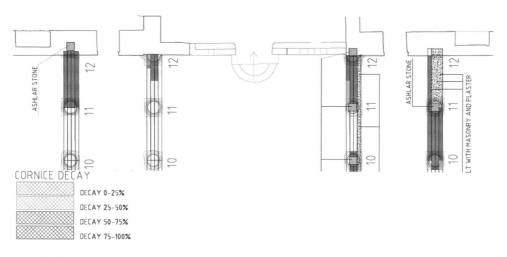

Figure 8.3.34 Examples of decay in lintel cornices.

Figure 8.3.35 X-ray equipment.

Some X-ray inspections were also carried out by using portable X-ray equipment in some parts of the architraves in order to detect possible metallic elements inside the beams or other relevant technical features not visible from the outside. Because of a potential danger to people, such analyses were performed during the closing time of the church (Figures 8.3.35, 8.3.36).

The following main features were detected (Figures 8.3.37, 8.3.38):

- internal wooden (hardwood) square chocks placed vertically and horizontally at different positions along the architrave, without nails

Figure 8.3.36 X-ray equipment.

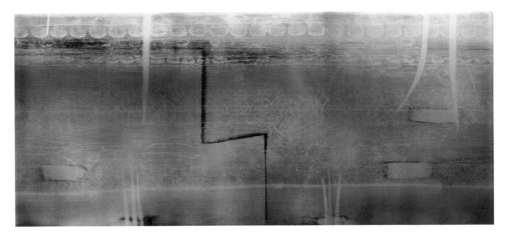

Figure 8.3.37 X-ray view: internal joint, nails and chocks.

- wooden (hardwood) nailed square chocks placed horizontally at the extrados and intra-dos of the beams and at the extrados of the cornice elements; one vertical or inclined nail per element was used
- nails between upper (central) elements and lower elements
- internal end half dovetail joint in the central beam above each column (not only above the third, sixth and ninth columns as for the external beams); when the central beam is composed by two elements the end joints are different: halved-scarf joint in the upper element and mortise-and-tenon joint in the lower element
- nailed iron staples where the elements had been replaced and the beams consolidated.

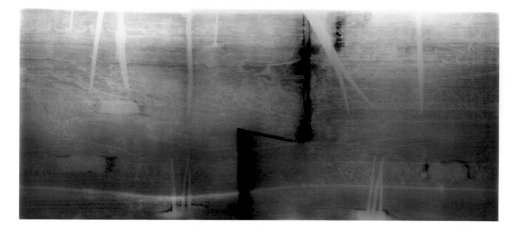

Figure 8.3.38 X-ray view: internal joint, nails and chocks.

It must be noted that, when looking at X-ray images, one should be aware of some visual defects, like distortions, due to the non-parallel rays' projection which could lead to false or wrong interpretations.

X-ray analysis provided a better understanding of the geometry of the composite beams especially with respect to some internal connections, like nails, non-visible end joints and non-visible wooden chocks. This analysis also confirmed the hypothesis that the composite beams were definitely assembled on the ground before being placed above the columns. Finally, no (metal) connectors were detected between composite beams and column capitals.

Moreover, wooden chocks, both nailed and not nailed, except for those placed at the interface between column capitals and beams (not detectable by other used non-destructive techniques), were detected also by means of other non-destructive techniques (drilling tests, thin metal blade gauge). Sometimes nails and internal end joints (like those between central beams) were also detected by drilling tests.

8.4   The new interventions

To the aim of the present section, a distinction was made between architraves, representing the whole series of wooden beams above each colonnade, and lintel, representing the single beam resting on two consecutive columns. Diagnostic surveys determined that most lintels were in a good state of preservation, with only a few of them having serious and extensive problems of fungal decay (e.g. Figure 8.1.23). Thus, for these elements static verifications were carried out to determine the need of interventions or even their complete replacement.

Elements were verified according to their residual strengths (ultimate limit states), whereas verifications according to deformability (serviceability limit states) were omitted. In fact, architraves are very old (they are dated back to the seventh century A.D.) and thus it can be assumed that deformations at infinite time have already occurred and are irreversible. To carry out structural verifications, lintels have been modeled as beams simply resting on two consecutive columns (although actually they rest on three columns – see Section 8.1 – probably for safety reasons) and subject to a vertical load distributed with parabolic law.

Moreover, since architraves have transversal section composed of several elements, some further simplifying hypotheses were made:

1   The contribution of the wooden planks forming the cornice was neglected in the calculation of the geometric and inertial characteristics;
2   The geometrical and inertial characteristics were given by the sum of the characteristics of the three beams constituting the lintels (in fact, they are put side by side and have equal height). In some cases, the central beam consists of two overlapping elements, and this occurrence was suitably modeled;
3   The decay quantified by the diagnostic survey was referred to the entire section as the sum of the decay in each beam;
4   To be on the safe side, the decay was considered as always positioned at the extrados: in fact, in this condition the section modulus is minimized and the bending stress maximized.

This way, the residual resisting section was calculated for each lintel.

Mechanical characteristics of wood were determined according to UNI 11119 [4], as described in Section 3.6, whereas for masonry and columns the values reported in the Final Report [1] were adopted.

To calculate the stresses acting on each lintel, the following loads were considered:

*   the self-weight;
*   the weight of the wooden planks forming the cornice;
*   the weight of the roof (structure and covering layers);
*   the weight of the wall;
*   the live load due to snow.

Considering the specific conditions, the load due to wind was obviously not considered.

Moreover, it was assumed that the relief arch above each lintel withstands the weight of the overlying structure, which is transferred to the side columns in both the nave and the aisles. Instead, the lintel only withstands the weight of the masonry filling occupying the space between the intrados of the small relief arch and the extrados of the wood lintel.

Once the geometry, the material characteristics and the loads applied to each lintel are known, it is possible to calculate for it the minimum safety section for which the beam can be considered safe. This section corresponds to the maximum admissible decayed section for which the beam is still verified; this also corresponds to identify the maximum decay class for which the lintel is no longer safe. Similar considerations can be made about the contact areas between architrave and stone arch or column capital.

For the nave lintels, a section reduction of up to 78% was calculated as admissible; therefore, consolidation was required only for the elements classified in the decay class 4 (75–100%).

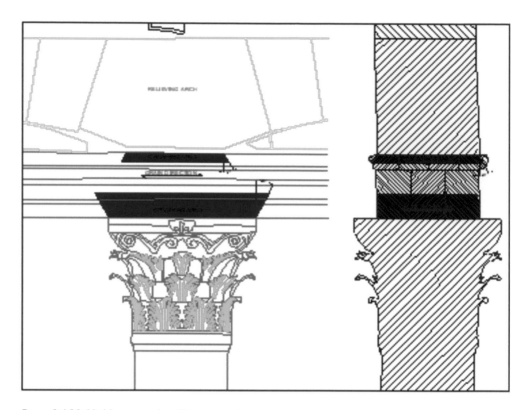

Figure 8.4.39 Highly stressed architrave portions.

For the lintels in corners, a section reduction of up to 50% was admissible, and therefore consolidation was required for the elements classified in the decay classes 3 (50–75%) and 4 (75–100%).

The architrave portions supporting the arches are very delicate and subject to high stresses (Figure 8.4.39). Surveys evidenced that also in past times some supports were repaired with the inclusion of stone cubes below the arch in place of decayed wood parts (Figures 8.3.29, 8.3.30). Moreover, stresses were different for the columns of the nave (which support the masonry walls and the nave covering) and those of the aisles, which support only the covering. Thus, in aisle lintels the classification in the decay class 4 (75–100%) induced to consolidation, whereas in nave lintels the interventions were needed for the decay classes 3 and 4 (50–100%).

All interventions affected lintels situated in both the nave and the aisles close to the central area (Figures 8.3.33, 8.3.34).

The interventions carried out on wood architraves were inspired by two main criteria:

- preserving (or recovering wherever it was changed) the static role of architraves;
- preserving as much as possible of the original material and decorations, therefore reducing replacements to the strictly necessary.

These interventions can be grouped in three types:

- Total replacement of the decayed element. This solution was adopted in the case of severe and extended decay and involved relatively small elements;
- Partial replacement of the decayed element, in the case of severe and not very extended decay;
- localized repairs, carried out when limited decay was present or when severe wood defects were found in non-decayed elements.

In total replacements, wood of the same species and of the same section as the original one was used; the resistant class was chosen equal to or better than the original one, corresponding to at least class 3 according to UNI 11119 [4].

In partial replacements, the missing parts were substituted by wood planks glued together on-site between them (similar to laminated timber). Old wood planks were used in the interventions. Also in this case, verifications were carried out following the prescriptions of Eurocode 5 (EN 1995–1–1:2008)[5], the Italian Technical Standards [6][7] and the Technical Document CNR-DT 206/2007 [8]. In calculations the Finite Element software PRO_SAP developed by 2si of Ferrara [9] was adopted.

As an example of total replacements, the case of the lintel LNS_T2.11 (lintel 11, resting on columns 11 and 12 of the south colonnade of the nave) is reported. The beam positioned at south had already been repaired in the past, whereas the north and central beams, as well as the wood cornice, date back to the Justinian's period. The old intervention consisted in substituting the south beam with an oak wood element and inserting a stone cube at both bases of the arch, in columns 11 and 12. The diagnostic survey showed that the replaced oak beam was completely decayed (Figure 8.4.40). Thus, an old oak beam of the same section as the original one was used to substitute the decayed element (Figure 8.4.41).

As an example of partial replacements, the case of lintel LNS_T1.10 (lintel 10, resting on columns 10 and 11 of the central colonnade of the south aisle) is described. This element had also been repaired in the past with the insertion of a stone cube at the arch base in column 11 (Figure 8.4.46). The diagnostic survey showed that the three beams and the cornice were from the time of Justinian (and hence original), but they were severely and extensively decayed (Figure 8.4.42), and verifications evidenced the elements were not safe.

Thus, it was decided to consolidate the lintels in the following way (Figures 8.4.43–8.4.47):

- The damaged parts were removed only internally, leaving unaltered the outer parts with the original decorations and engravings from the time of Justinian;
- These external "walls" were consolidated following the methodologies usually adopted for artworks (and hence of not-structural type);
- Some old wood planks were inserted in the empty internal parts and glued together (Figure 8.4.47). The new laminated element has variable section to follow the contour of the volume removed for decay, and of course it plays a structural role in the repaired lintel (Figures 8.4.45, 8.4.46);
- A new stone cube was inserted at the arch base in column 10 to replace the degraded wood of the lintel (Figure 8.4.46).

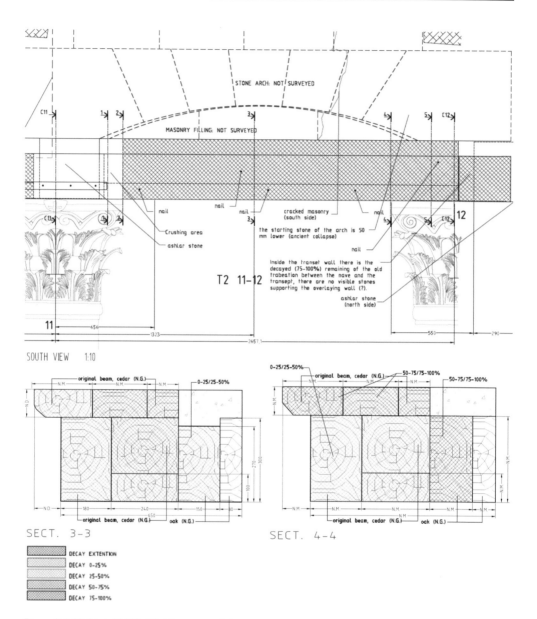

Figure 8.4.40 Lintel LNS_T2.11: survey of decay.

It is worth noting that the resisting section considered in the structural verification was the minimum one contained in the variable section of the laminated beam.

In general, before proceeding with the actual consolidation, the plaster was removed whenever it was present on both sides. Also the masonry filling between stone arch and architrave was removed. Cornices and upper parts of the architrave were also removed and restored if their state of decay was considered beyond the admissible threshold.

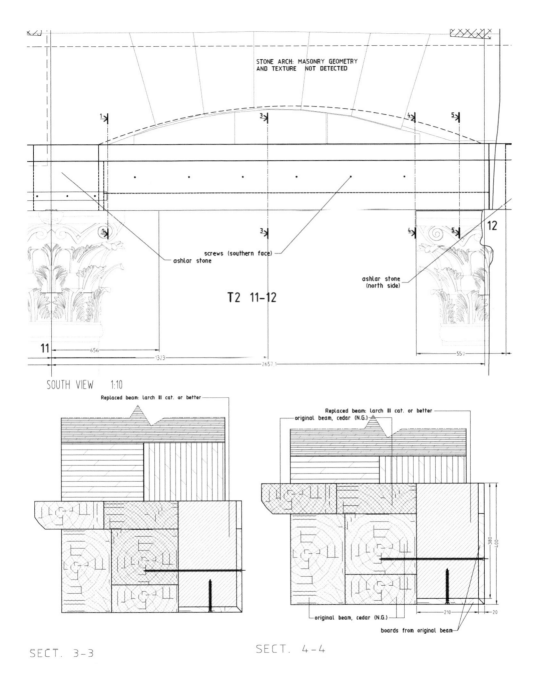

STONE ARCH: MASONRY GEOMETRY
AND TEXTURE NOT DETECTED

screws (southern face)
ashlar stone

ashlar stone
(north side)

T2 11-12

12

11

SOUTH VIEW   1:10

Replaced beam: Larch III cat. or better

SECT. 3-3

Replaced beam: Larch III cat. or better
original beam, cedar (N.G.)

original beam, cedar (N.G.)

boards from original beam

SECT. 4-4

Figure 8.4.41 Shop drawings of the intervention on LNS_T2.11.

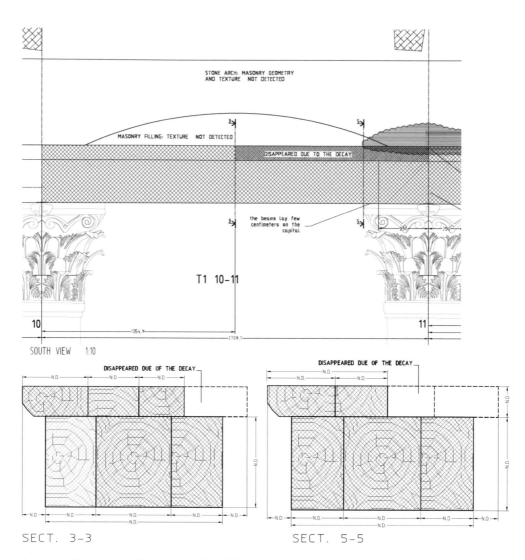

STONE ARCH: MASONRY GEOMETRY
AND TEXTURE NOT DETECTED

MASONRY FILLING: TEXTURE NOT DETECTED

DISAPPEARED DUE TO THE DECAY

the beams lay few
centimeters on the
capital

T1 10-11

10

11

SOUTH VIEW   1:10

DISAPPEARED DUE OF THE DECAY

SECT. 3-3

DISAPPEARED DUE OF THE DECAY

SECT. 5-5

Figure 8.4.42 Survey of decay on lintel LNS T1.10.

When the external parts had to be consolidated, a thin layer of Paraloid B72 was spread on the inner side with a brush and then covered with a hemp canvas fixed to the support with a compound of vinyl resin, water, red pozzolana and sawdust (Figure 8.4.48). Of course, this procedure was followed just to keep the original (and decorated) external part well in contact with the load-resisting element, and not for structural-type interventions.

Moreover, in several cases the connection between the new insertions and the outer parts was guaranteed by horizontal and vertical screws (Figure 8.4.49).

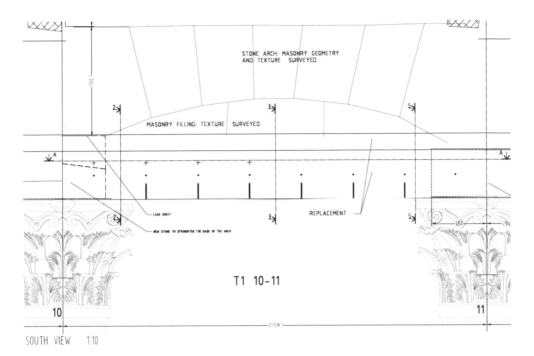

Figure 8.4.43 Front view of the intervention on lintel LNS T1.10.

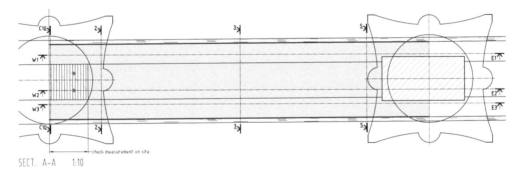

Figure 8.4.44 Plan view of the intervention on lintel LNS T1.10.

It is also worth evidencing that the interventions allowed confirming the geometric description based on diagnostic surveys, carried out with visual and instrumental (resistographic and X-ray analyses) inspections (Figures 8.4.50–8.4.53). Moreover, during the intervention on lintel LNS_T2.11, it was found that the original beam positioned centrally was carved although it was an internal and not visible element (Figure 8.4.54). This might let

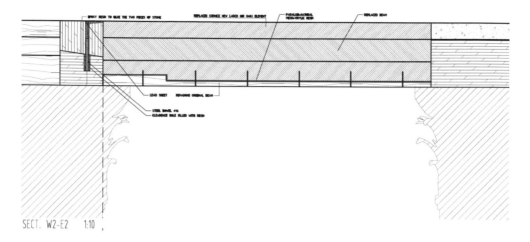

Figure 8.4.45 Representative longitudinal section of the intervention on lintel LNS TI.I0.

suppose that in the reconstruction of the basilica in Justinian's period, some elements were reused from the remains of the previous basilica of Constantine.

In the end, 52 segments of the wooden architraves with a total length of 154 m were renovated. The works included also the consolidation and protection of all decorative outer surfaces.

Considering that the walls above the wooden architraves rest on stone arches, the stability of these arches was also checked. Two verifications were carried out using the Arco software [10]: one for the arch stability and one for the masonry strength. The first one was performed with design loads lower than the characteristic ones ($\gamma = 0.9$), because this condition was the most penalizing with respect to stability; the second one considered design loads higher than the characteristic ones ($\gamma = 1.3$) because this condition was the most penalizing with respect to the material strength. Both verifications evidenced that the stone arches were in safe condition: the thrust line was contained within the arches' thickness, with at least 78% of the section compressed; the maximum compressive stress reached in the arches section was 1.62 daN/cm^2, well below the estimated limit value of 45.5 daN/cm^2.

Moreover, the thrusts of the arches resulted in all being balanced, with the exception of those of the initial and final spans. In fact, since in the past consolidation works were carried out in the last lintels of the whole architraves (east side; see Section 8.3), consisting in the insertion of stone cubes below the arch support (in place of wood), currently the wooden lintels, and therefore the architrave, do not work as tie beams anymore (Figure 8.4.55). Probably also for this reason, a wall was built at the east ends of the aisles, and two buttresses were inserted to the west ends of the nave colonnades. However, the stability verification carried out on the thrusts of the end arches was positive, confirming the effectiveness of the interventions carried out in past times.

The various works were completed in October 2016.

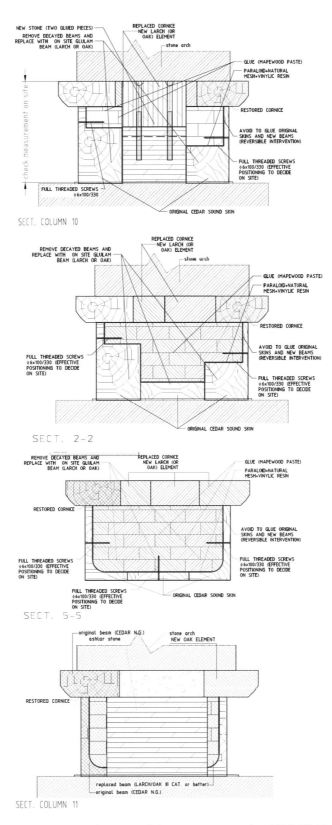

Figure 8.4.46 Representative transversal section of the intervention on lintel LNS T1.10.

Figure 8.4.47 An example of partial replacements of the internal decayed beams of lintels; a stone cube in the end.

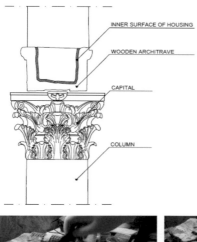

INNER SURFACE OF HOUSING

WOODEN ARCHITRAVE

CAPITAL

COLUMN

Figure 8.4.48 Phases of the consolidation of the lintel external parts.

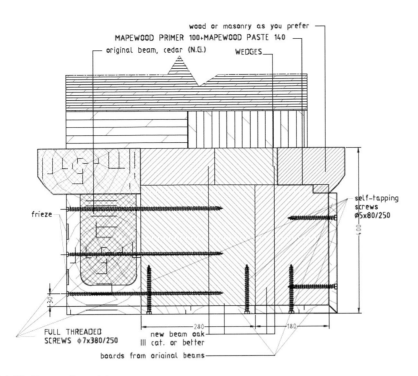

Figure 8.1.19 Connection with screws.

Figure 8.4.50 Overlapped inner beams.

Figure 8.4.51 Half-lap joint and chock.

Figure 8.4.52 Horizontal chock.

Figure 8.4.53 Chock detail.

Figure 8.4.54 Inner carved element.

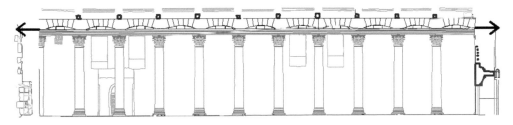

Figure 8.4.55 Architrave with end horizontal thrusts balanced by added buttresses and walls.

8.5 The problem of the corners

Particular attention must be deserved to the two corners, called Armenian and Orthodox corners after the 1852 *status quo* that placed them under the direct control of the Armenian (north corner) and the Orthodox (south corner) Communities. Placed to the north-east and south-east of the church respectively, at the intersection between transept and presbytery, as marked in red in Figure 8.5.56, they are symmetrical with respect to the longitudinal axis of the church and differ only in some details that are negligible from a structural point of view.

Each corner includes five stone columns (numbered in Figure 8.5.57), as tall as all the others of the church and surmounted by wooden lintels identical to those of the nave and the aisles.

Also here, four relieving arches spring directly from the architraves with a span equal to the distance between two columns. They support the walls above them, as high as the walls of the nave, and part of the load transferred by the half trusses that cover the more external corner areas (Figure 8.5.58).

At the top level of the walls of each corner, four windows opened, two on each side. Some of them, especially those closest to the corners, were buffered, probably after the seismic events occurred in the first half of the 19th century (Figure 8.5.59).

In particular, the investigations carried out in the architraves of these two corners highlighted a structural problem already envisaged in 1932 by Harvey during his archaeological campaign [11]. In fact, the central columns of both corners (column n. 3 in Figure 8.5.57 and its correspondent in the opposite corner) show some weakness planes differently oriented, which have always raised concerns since their first detection at the beginning of the last century (Figures 8.5.60–8.5.63).

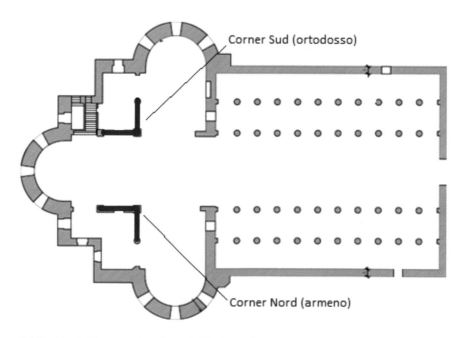

Figure 8.5.56 North (Armenian) and south (Orthodox) corners.

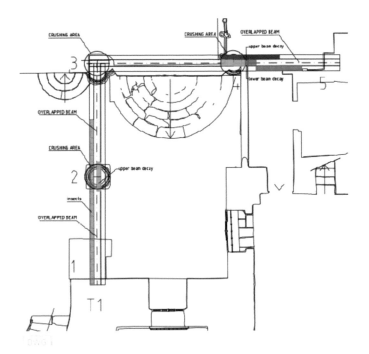

Figure 8.5.57 South (Orthodox) corner.

Figure 8.5.58 Half trusses in the corner areas.

The recent restoration works carried out in the church, in particular in the architraves of the corners, offered the opportunity to analyze in detail such a problem.

It was thought at first that such weakness planes (it is not clear yet if they correspond to actual cracks) might be the consequence of high thrusts transferred from the relieving

Figure 8.5.59 Buffered window in a corner.

Figure 8.5.60 Weakness plane in the north column.

Figure 8.5.61 Weakness plane in the north column.

Figure 8.5.62 Weakness plane in the south column.

Figure 8.5.63 Weakness plane in the north column.

Figure 8.5.64 Physical decay in a lintel.

Figure 8.5.65 Architrave interrupted by a stone nut.

arches to the lintels and from these to the capitals of the columns through the only friction at the interfaces between these different architectural elements, as no particular connection, like nails or iron rods, was found between masonry, architraves and capitals. Moreover, it is worth noting the architraves of these corners showed clear signs of physical decay (Figure 8.5.64), and they had lost their ability to work as tie beams after the replacement of wood with a stone nut in correspondence of some columns (Figure 8.5.65).

Analogous interventions had been carried out in the past also in other architraves of the church (see Section 8.3), always above columns, in replacement of decayed wooden parts, which, under the load transferred by overlying arches, probably appeared strongly crushed.

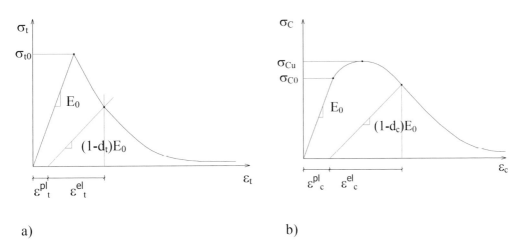

Figure 8.5.66 Mechanical behavior of masonry under uniaxial tension (a) and compression (b).

However, initial analyses carried out by assuming material linear elastic behaviors showed that the horizontal component of the result of the two thrusts coming from the two corner arches was too high to be absorbed by friction alone. It was therefore thought of a possible interpenetration of the masonry in the lintel and of the latter in some relief of the top surface of the capital, such as to provide an effective constraint to the relative sliding of one element above the other, but a more accurate survey of the interfaces excluded the existence of any interpenetration. Moreover, there were no signs of damage on the walls, such as to denounce some mechanism of collapse and, even if some cracks had occurred in the past, they had been covered over the centuries by thick layers of plaster and by the wall mosaics. Therefore, at present, despite those weakness planes surveyed in the corner columns, both corners seem to be quite stable. Nevertheless, pending further on-site inspections, some numerical simulations were performed in order to find a possible and reasonable justification of the present stability of the corners and, above all, to check the level of their seismic vulnerability.

Both corners were modeled by using heterogeneous approaches [12] with a distinction between stones and mortar joints. The choice of a heterogeneous model allowed taking into account the interactions among stones, the constructive features of the arches, the way in which windows were buffered, all issues that influence the actual behavior of the corners and that in no way could be taken into account by other kinds of approach.

The non-linear behavior of masonry was described by means of the Concrete Damage Plasticity (CDP) model available in the ABAQUS/Standard materials library [13–15]. The CDP model is a continuum plasticity-based damage model in which the main failure mechanisms are tensile cracking and compressive crushing. The model is able to assign different strengths, stiffness degradation and recovery effect terms both in tension and compression, as schematically shown in Figure 8.5.66.

The geometrical model is generated by Abaqus and consists of that part of the corner including one arch and a half on each side, three columns, one of which is at the intersection of nave and transept, architraves and walls and the existing openings (Figure 8.5.67).

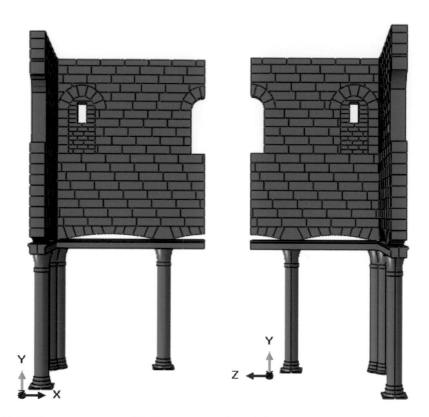

Figure 8.5.67 Heterogeneous model of one corner – lateral views.

The filling between arches and lintels has not been considered because it is made of scarcely consistent material. Moreover, the bases of the columns are clamped, and symmetry in X and Z directions is considered, with X and Z parallel to nave and transept, respectively.

The arches are simply supported by the lintels, as well as the lintels, modeled as linear elastic materials, are simply supported by the columns. Therefore, each component can slide on the other, and the interaction at the interface is defined on the basis of suitable values of the static and dynamic friction coefficients. Load conditions and material parameters are defined with reference to the Italian Technical Standards [6][7] and the data surveyed on-site. The loads are incremented by apposite multipliers in order to evaluate the evolution of internal damage and crack pattern.

The joints are modeled with 3-D elements, 1 cm thick, although the actual joints have a smaller thickness. The assembly of the stones is consistent with the data of the survey. Where the walls are covered either by plaster or by mosaics an assembly of stones was hypothesized with vertical joints staggered by 15 cm in order to assume the most unfavorable configuration. Also the stones of the relieving arches were faithfully reproduced (Figure 8.5.67). The columns were modeled as unique stone blocks as they are in reality. It is worth noting that in the concrete damage plasticity material model used it was necessary to modify the strength characteristics in order to simulate the different mechanical behavior of mortar and stone, so as to obtain more realistic diagrams for the plastic behavior in tension and compression.

The numerical analyses were carried out with three increasing load conditions starting from the lowest, assumed as basic load condition. The other two are obtained from the first one by using multipliers 2 and 4, with multiplier 2 corresponding to the ultimate limit state (ULS) at the safety limit considered in the Standards.

Even the static and dynamic friction coefficients between lintels and stones vary between basic values provided by the technical literature and infinite values corresponding to possible states of interpenetration at the interfaces masonry–lintel or capital–lintel.

Two sensitivity analyses were carried out, one with constant basic friction and increasing loads, the latter with basic constant loads and increasing friction up to an infinite value.

As is obvious, the horizontal thrust components at the corners increase in both cases; on the contrary, the maximum vertical displacements of the arches increase considerably only with the load (from 5.4 mm to nearly 23 mm in the arch along the nave and from 5 mm to nearly 16 mm in the arch along the transept), whereas they are not very affected by changes in friction values. It is worth noting that no filling was assumed between lintels and arches. This assumption can be justified, as mentioned before, by the presence of a low-quality filling material which would not be able to contrast possible vertical displacements of the arch (Figure 8.5.68).

The concrete damaged plasticity constitutive law chosen to model the material behavior of masonry allows identifying the zones affected by damage due to tension and therefore by cracks, as shown in Figure 8.5.69. The areas with lower strength or affected by cracks start from the arches in proximity of abutments and key stones (see the behaviors of these arches in [12]) and move upwards, as the loads increase, converging towards the common edge.

It seems therefore that new and higher relieving arches are formed in the walls above the actual ones so as to diminish considerably the thrusts on the central column. It is also possible to say that the loose filling material is compressed and compacted in consequence of the arch lowering, so as to transfer to the lintel the weight of that part of the wall now defined under these new arches. Therefore, the lintels would not be subject to tension only but also to vertical loads and bending moments.

Figure 8.5.68 Filling between arch and lintel.

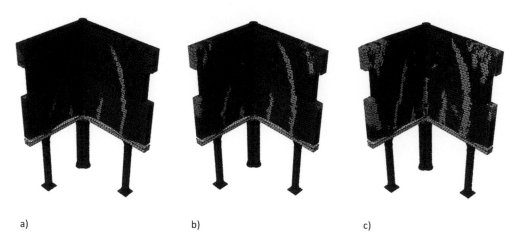

a)   b)   c)

Figure 8.5.69 Crack pattern due to (a) basic loads, (b) loads multiplied by 2, (c) loads multiplied by 4.

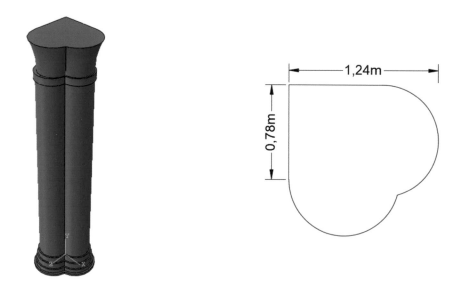

Figure 8.5.70 Central column of the corner.

This might be a possible explanation of the present stability of these corners in the presence of vertical loads. However, this stability would be seriously endangered by the occurrence of seismic actions that were not taken into account in the verifications.

As for the central column (Figure 8.5.70), having a bigger cross section than other columns, the analysis carried out with basic friction coefficients and basic loads shows that it moves towards the central area, approximately in the same direction as the horizontal component of the thrust resultant.

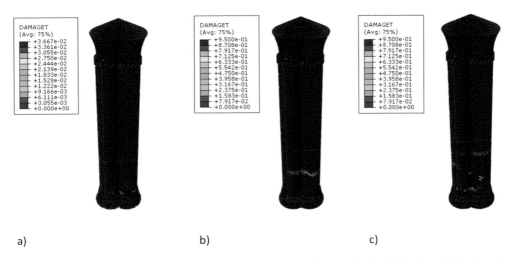

Figure 8.5.71 Damage Index in Tension with increasing loads (a) basic loads, (b) loads multiplied by 2, (c) loads multiplied by 4.

With constant basic friction and loads multiplied by 2 the horizontal displacement of the top of the column is about 5 mm. It increases considerably by increasing the loads, but it remains almost constant with increasing values of friction. That means that, with the assumed mechanical model and the state of damage obtained, the particular conditions at the interfaces do not affect considerably the behavior of the column, which instead depends prevalently on the load conditions. Moreover, the horizontal displacement mentioned above and computed with ULS load condition is still a fairly low value and such as not to jeopardize the stability of the corners, but nevertheless, it already corresponds to an internal damage, as shown in Figure 8.5.71, which, in the presence of additional loads, like seismic ones, might activate collapse mechanisms.

In conclusion, the results obtained by these analyses can justify the present stability of the corners, despite the presence of some damage in the columns, but only if vertical loads are assumed. But Bethlehem is in a seismic area, and additional seismic actions certainly would endanger the stability of the corners.

8.6 Maintenance

For general considerations on the maintenance plan of structures, the reader is referred to Section 3.10.

This section deals with the wooden elements and the metal and non-metal connection elements that compose the architraves above the colonnades.

8.6.1 Wooden elements

User manual

They are all those elements that contribute to define the wooden section of the architraves that surmount all the colonnades of the church. They are made of solid cedar and date back to the time of Justinian. The joints between the various elements forming the generic cross section are made of forged iron nails and wooden chocks. The identification of all connections, hidden inside the architraves, was only possible through a careful instrumental survey, which also involved the use of X-ray techniques. Restoration and/or consolidation interventions were carried out by disassembling the beams, without damaging the visible parts, and by rebuilding them with laminated wood assembled on-site. The original nails were completely reused after being straightened and checked in their quality and integrity.

The correct use of these elements requires the maintenance of operating loads within the project limits and the non-alteration of the surface beams over time.

Maintenance manual

> *Minimum level of performance*: resistance to design stresses and preservation of the original surfaces and decorations.
>
> *Abnormalities that can be found*: attack by insects and fungi due to percolation of water and permanence of high levels of humidity for a long time. During the diagnosis and restoration phase, the presence of decay (now completely over) due to termites was detected, although not directly on the architraves. The presence of termites is particularly dangerous because the damage caused is generally severe and difficult to detect. The location of the architraves should ensure adequate protection against the attack by termites. However, it is necessary to control carefully the presence of these specific pests.
>
> *Controls*: a first check is carried out through a visual inspection aimed at identifying particular signs of degradation. Later, if decay has been actually detected, a more in-depth analysis must be performed, including an instrumental survey.
>
> *Frequency of checks*: every year the visual inspection, when necessary other specific analyses.
>
> *Carrying out checks*: the visual inspection can be performed by the trained user, whereas the other analyses must be carried out by specialized personnel.
>
> *Interventions*: they must be carried out by specialized personnel, depending on the problems encountered. They include protective treatments of elements slightly attacked by insects, in order to prevent the infestation spread to healthy material and hence the damage progression in the attacked wood. Use of preservatives as impregnating (and not filmable) products.
>
> *Frequency of the interventions*: whenever it is necessary. Preservatives against insects should be applied as a minimum after three years the first time, after five years the second time and every 10 years the subsequent times.
>
> *Carrying out of the interventions*: they must be performed by specialized personnel.

8.6.2   Prostheses

User manual

They are the parts (internal and not visible from the outside) made of laminated wood, assembled on-site, which replace the portions irreversibly damaged (Figure 8.6.72). Their

a)   b)

Figure 8.6.72 (a) Two component prosthesis, (b) three component prosthesis

durability is guaranteed by the use of naturally durable wood, that is, naturally resistant to fungi and insects.

The correct use of these elements requires the maintenance of operating loads within the project limits and their non-alteration over time.

Maintenance manual

Minimum level of performance: resistance to design stresses.

Abnormalities that can be found: attack by insects and fungi due to percolation of water and permanence of high levels of humidity for a long time; delamination and/or detachments of the laminated wood. Since these elements are invisible from the outside, these possible anomalies cannot be directly detected. During the visual inspection, particular attention must be paid to the identification of phenomena such as local deformations, excessive deformations, incipient failures, differential displacements between the various elements forming the resistant section of the beams etc., which may indicate a possible problem concerning the prosthesis.

Controls: they are performed through a visual inspection aimed at identifying the possible presence of internal damage of the prosthesis. In case of substantial doubt, a suitable instrumental investigation is necessary, carried out by highly specialized experts, in order to verify the actual occurrence of the damage.

Frequency of checks: every year the visual inspection, when necessary other specific analyses.

Carrying out checks: they must be performed by specialized technicians.

Interventions: if there are anomalies in the prosthesis, both numerical verifications/simulations and interventions must be carried out by highly specialized experts.

Frequency of the interventions: whenever they are necessary.

Carrying out of the interventions: they must be performed by specialized personnel, on the basis of technical reports and shop-drawings provided by highly specialized experts.

References

[1]   Ferrara Consortium (2011) *Final Report*.

[2]   Bernabei, M. & Bontadi, J. (2012) Dendrochronological analysis of timber structures of the Church of the Nativity in Bethlehem. *Journal of Cultural Heritage*, 13, Elsevier, e5–e26.

[3]   Bacci, M. (2017) *The Mystic Cave: A History of the Nativity Church in Bethlehem*. Rome, Brno, Viella/Masaryk University Press.

[4]   UNI 11119:2004 *Beni culturali – Manufatti lignei – Strutture portanti degli edifici – Ispezione in situ per la diagnosi degli elementi in opera* (*Wooden artefacts: Load bearing structures: on-site inspections for the diagnosis of timber members*).

[5]   EN 1995–1–1:2008 (Eurocode 5) *Design of timber structures*. Part. 1–1: General-Common rules and rules for buildings.

[6]   Circolare 617/2009 *Istruzioni per l'applicazione delle nuove Norme Tecniche per le Costruzioni di cui al decreto ministeriale*, 14 Gennaio 2008.

[7]   Decreto Ministeriale 14 gennaio 2008, Norme Tecniche per le Costruzioni.

[8]   Technical Document CNR DT 206/2007 *Istruzioni per la Progettazione, l'Esecuzione ed il Controllo delle Strutture di Legno*.

[9]   PRO_SAP *Professional Structural Analysis Program*, 2SI, Software e Servizi per l'Ingegneria, Ferrara, Italy Build 2015.12.173 (ver. 16.0.0).

[10]   ARCO Software ARCO by Gelfi, P., University of Brescia, Italy, ver.1.2 07/04/2008.

[11]   Harvey, W. et al. (1910) *The Church of the Nativity at Bethlehem*. London, B.T. Batsford.

[12]   Alessandri, C., Del Balzo, M., Milani, G. & Valente, M. (2018) A structural paradox in the church of the nativity: FE static analyses. *10th IMC Conference Proceedings*. pp. 2532–2554.

[13]   Lee, J., Fenves, G.L. (1998) Plastic-damage model for cyclic loading of concrete structures. *Journal of Engineering Mechanics*, 124, 892–900.

[14]   Lubliner, J., Oliver, J., Oller, S. & Oñate, E. (1989) A plastic-damage model for concrete. *International Journal of Solids and Structures*, 25, 299–326.

[15]   ABAQUS®, Theory Manual, Version 6.14.

Chapter 9

The narthex

9.1 Archaeological analysis and preliminary structural survey

C. Alessandri, G. A. Fichera and M. Martinelli

The main façade of the basilica is probably the one that has undergone over the centuries the greatest number of changes and adjustments that distorted the original appearance, transforming the sumptuous basilica, as it appeared to a visitor of the mid-sixth century, into the impregnable fortress we still observe today. The archaeological analysis of the elevations made it possible to attribute the construction of the supporting structure of the present façade to the Justinian era, together with the reconstruction of the entire basilica [1][2]. In the same period the present narthex replaced a previous entrance portico that was partially brought back to light by the archaeological excavations carried out in the 1920s during the British Mandate in Palestine by R. Hamilton [3]. The archaeological excavations carried out by the English and the subsequent reworking of Vincent [4] and Bagatti [5] show how the façade of the original church was set back eastwards, a few meters from the current one, along with the entrance hall that opened onto the large square portico (Figure 9.1.1).

The new entrance to the basilica, entirely rebuilt in the sixth century, was characterized by a large central portal with a moulded lintel, protruding from the façade plane, set on large corbels decorated with scrolls. The lintel was surmounted by a lowered or relieving arch, still partially visible (Figure 9.1.2).

On both sides of the portal there were two doors with similar configuration but smaller in size and two small windows (Figure 9.1.3).

What remains of the original masonry is characterized, similarly to the other perimeter walls of the basilica, by a construction technique that uses medium-large blocks of malaki stone, perfectly squared and finished on surface, placed on horizontal and parallel rows with very thin and regular mortar joints. No doubt anyone who entered the narthex was astonished by the façade, with its elegant alternation of full and empty and the finesse of the decorative details. Unfortunately, all of this is difficult to perceive today due to the closure of almost all the entrances, most likely after it was conquered by the Crusaders who upon their arrival, at the end of the eleventh century, transformed the basilica, although still in good conditions, into a military fortress.

To the same period dates back the construction of a large building located on the south side of the basilica, now the seat of the Armenian Community, which has largely covered the south portal, as well as the pointed arch built inside the original portal (Figure 9.1.4).

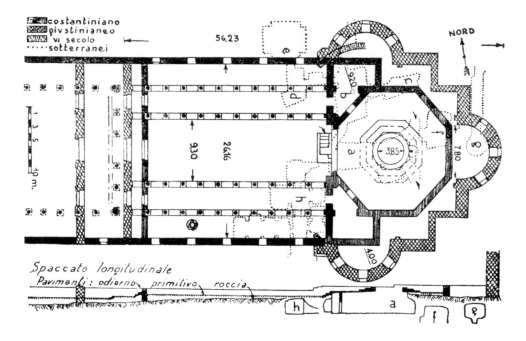

Figure 9.1.1 Plan of the basilica of Constantine.

Figure 9.1.2 The main portal with subsequent modifications.

WEST ELEVATION

Figure 9.1.3 Layout of the lateral doors.

Figure 9.1.4 The Armenian complex on the right-hand side.

a)   b)

Figure 9.1.5 Protruding stones in the north corner of the narthex (a) and in the transept (b).

Some centuries later, under the Ottoman rule (16th century), the central portal underwent a further reduction due to the insertion of a massive stone lintel which transformed the entrance into a small access which took the name of "Humility Door", as it forced visitors to bow to access the basilica. This intervention was carried out to prevent the soldiers from entering the church with horses, camels and other animals. Meanwhile, the nave had become a sort of separate space, reserved for Muslim prayer and for secular activities, such as the justice administration.

It is interesting to note how the northern corner is characterized by the presence of large stones that protrude from the vertical of the corner itself (Figure 9.1.5 a), defining a "*dentelli*" system similar to that identified at the junctions between the apses and the walls of the aisles, with purely decorative function (Figure 9.1.5 b).

Two of the internal walls, the one to the north adjacent to Saint Helena's Chapel and the one to the south delimiting the access area to the Armenian courtyard, also seem to confirm Bagatti's hypothesis of two bell towers occupying the north and south corners of the narthex. In the façade above the narthex, the inclined profiles of the original pitches of the roof of the nave are clearly visible, upon which the masonry was subsequently raised (Figure 9.1.6).

As in its original configuration, the narthex is today divided into five bays surmounted by five masonry cross vaults that replaced the previous roof in the Crusader time. Unfortunately, however, today it is not possible to enjoy, as in the past, the perception of a unitary space because of the presence of the two internal walls, orthogonal to the façade and belonging to the base of the bell towers mentioned above, and a further wall, also orthogonal to the façade, built fairly recently to delimit a space now reserved for the surveillance service (Figure 9.1.7).

The structure of the narthex is today strongly deformed. The façade wall of the narthex exhibits a rotation towards Manger Square more pronounced in the middle and starting approximately from the lintel of the Humility Door, about 1 meter above the ground.

Figure 9.1.6 Inclined profile if the original roof pitches.

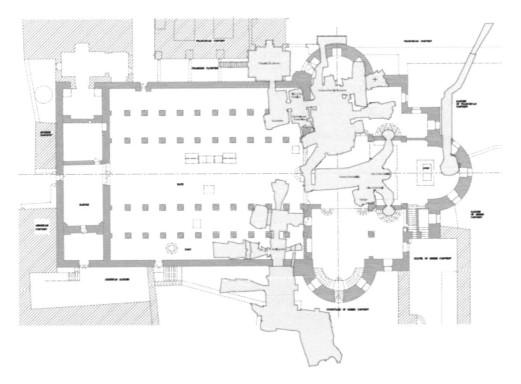

Figure 9.1.7 Plan of the church and narthex.

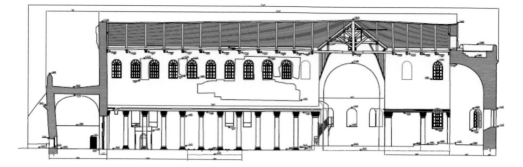

Figure 9.1.8 Outward rotation of the narthex façade.

Figure 9.1.9 The original buttress in an old drawing.

In consequence of this rotation, the façade wall exhibits a maximum horizontal displacement at the top, approximately in the middle of the free part of the wall, of about 40 cm (Figure 9.1.8).

It is a very high value if compared with the height of the wall, about 10 meters measured at the top. If some cracks occurred in the façade in the past, these are now closed after some local interventions of cleaning and consolidation made over the centuries. This is a clear evidence that the damage evolution is now over and that it was probably over even before the addition in 1779 of an external buttress, facing Manger Square, and built with the aim of stopping the rotation of the façade towards the square (Figure 9.1.9). This buttress covered the north portal and part of the central one.

The comparison of the current situation with the oldest representations of the basilica shows that, during the XIX century, the buttress was partially dismantled to allow the reopening of one of the original windows and to illuminate the guardroom (Figure 9.1.10).

Also the façade of the church, opposite to the one of the narthex, exhibits a light out-of-plane deformation which starts from the roof of the narthex and achieves 10 cm at the top of the tympanum.

During the English Protectorate a wooden scaffolding was built inside the narthex to support the central vault from the intrados. There are no documents that explain whether this scaffolding was built following an extraordinary event or as a simple precaution, given the considerable state of deformation of the vaults and the worrying inclination of the façade of the narthex (Figure 9.1.11).

Figure 9.1.10 The current buttress after the removal of the central part.

Figure 9.1.11 Scaffolding built during the British Mandate.

Figure 9.1.12 Spots of moisture on the vaults.

It was urgent to remove the existing scaffolding and to carry out an appropriate restoration that could remedy the damage occurred over time and ensure the stability of the whole structure even in the presence of seismic actions. Structural damage was also the cause of considerable infiltrations of water inside with severe degradation of the plaster and masonry (Figure 9.1.12).

In accordance with the list of priority decided by the Presidential Committee, the restoration of the narthex was awarded to the contractor on March 28, 2014. Since the beginning, according to the main criteria that guided the restoration of the whole church, there was a unanimous consensus in preserving the existing vaults despite their precarious material and structural conditions.

In the first place, an external temporary roof and the necessary propping system inside the narthex (Figure 9.1.13) were set up in the two bays adjacent to the central one, which had already been propped in the last century. In fact, the crack pattern affected not only the central vault but, to a lesser extent, also the adjacent ones.

Then an excavation campaign started in August 2014 on the narthex terrace to verify the state of conservation of the three central vaults, here named vaults number 2, 3 and 4, in order to get information about times and methods of construction of such vaults.

The terrace was paved with tiles of local limestone, except in the central portion where in the past the slabs had been replaced by layers of cement and bitumen. All the tiles were numbered and documented in a photo-plan in order to be relocated at the end of the consolidation and restoration intervention (Figures 9.1.14, 9.1.15).

The excavation began with vault number 3, the central one, which presented major problems and it was subsequently extended to the adjacent vaults.

Figure 9.1.13 New internal scaffolding.

Figure 9.1.14 Tiles numbering.

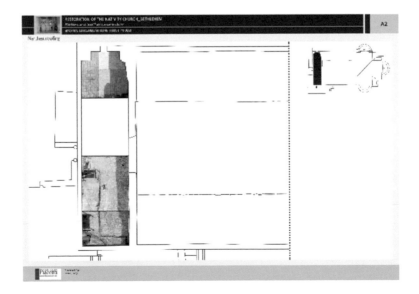

Figure 9.1.15 Photo-plan.

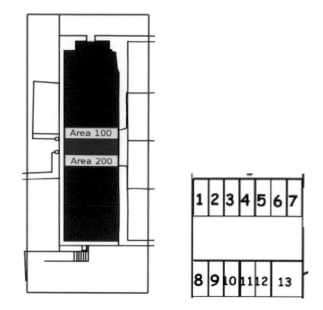

Figure 9.1.16 Reference grid on the central vault.

The removal of the covering material of the vaults was performed following a strati-graphic method of analysis, in order to allow an analytical reconstruction of the interventions made over the centuries.

On the extrados of the central vault a grid was drawn so as to define 13 samples (Figure 9.1.16): 7 on the north side of the vault, forming the area 100, and 6 from 8 to 13 forming the area 200 on the south side.

The choice to work with samples of small dimensions depended on the need to allow a parallel and simultaneous control by a team of engineers of the critical issues that emerged on the extrados of the vault while the archaeological strata were removed. The samples were then referred to the adjacent macro stratigraphic areas, so as to unify the data reading and facilitate the understanding of the whole stratigraphy. On the contrary, the north and south vaults, less damaged than the other, were analyzed within macro areas.

The removal of the floor slabs brought to light, on a large part of the investigated area, a level of lime mortar, of a thickness of about 5 cm, fairly well preserved but endowed in the centre in the north–south direction and characterized in correspondence of the eastern edge by a series of rather deep fractures (Figure 9.1.17), signs of the structural collapses that probably led to the construction of the external buttress and then to the shoring of the central vault in the last century.

This mortar floor covered a preparation level consisting of well-placed stones in layers of lime and sand (Figure 9.1.18), below which the space at the four corners of the cross vaults had been filled with layers of sand and soil mixed with crushed stone.

The layers of soil returned numerous fragments of pottery, mosaic tiles some of which still adhered to mortar fragments, coins, animal bones, fragments of iron nails and even a piece

Figure 9.1.17 Lime mortar level.

Figure 9.1.18 Level of stones under the mortar floor.

of lead sheet. It is worth noting that some mosaic tiles are larger than those used for the wall mosaics and probably belonging to a floor level (Figure 9.1.19).\

A slightly different situation was identified in the southern side of the narthex, from about half of vault 3 and throughout vault 4, where the floor mortar level had been removed during a past restoration intervention to build a masonry structure directly above the extrados of

Figure 9.1.19 Archaeological finds.

Figure 9.1.20 Masonry structure above the south vaults.

the vaults (Figure 9.1.20). Even the ceramic finds confirm a later period for the restructuring intervention, compared to the one of the vaults' construction. In fact, they are ceramics that date back at least to the end of the 14th century. Moreover, the remains found in the filling of these vaults are mostly different from the ones of the northern side and consist of numerous fragments of tiles, floor and wall mosaics.

It is possible to hypothesize that during the restoration of the vaults the entire basilica underwent a general renovation that probably involved the wall mosaics, perhaps partly deteriorated, and the roof covering, which from a tile covering might have been turned into a lead covering.

The archaeological excavation and all the works carried out upon the central vaults of the narthex allowed advancing hypotheses of fundamental importance about the transformations that the basilica underwent over the centuries. First of all, they made it possible to affirm with certainty that the construction of the vaulted system dates back to the Crusader period, as proved by the ceramic finds under the mortar floor, dating back, approximately, to the end of the 12th century.

During the excavation it was also possible to reopen some holes that were used to drain the rainwater and that had been closed over the years (Figure 9.1.21).

From some traces found in the façade of the church and in the counter-façade of the narthex during the restoration works it was possible to deduce that the original roof structure of the narthex was a wooden roof, as in the majority of the early Christian basilicas with narthex. The structural details of this roof are obviously unclear, although the geometrical characteristics of the traces on the walls suggest a strong similarity with the trusses of the aisles. In fact, on the church façade, immediately below the level of the existing floor it was possible to see a set of rectangular holes 45 × 48 cm and 82 cm deep, distant from each

Figure 9.1.21 Original hole for water draining.

Figure 9.1.22 Bigger and smaller holes and recess on the façade of the church.

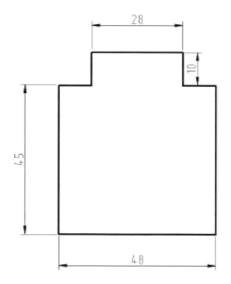

Figure 9.1.23 Dimensions of the bigger hole.

other 2,35–2,40 m, exactly like the holes that host the ends of the half trusses in the aisles (Figures 9.1.22, 9.1.23, 9.1.24). The space 10 × 28 cm above the major rectangular space was necessary to facilitate the insertion of the struts and guarantee a better ventilation all around the ends of the wooden elements.

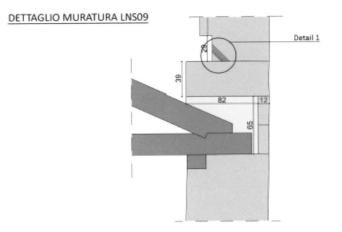

DETTAGLIO MURATURA LNS09

Figure 9.1.24 Cavity for truss ends of the aisles.

Figure 9.1.25 Small hole on the counter-façade of the narthex.

Immediately below these cavities (Figure 9.1.22) there is a rectilinear recess that suggests the presence in the past of a wooden lacing, as in the trusses of aisles and nave. Moreover, further below, it was possible to note some small holes at regular distances (Figure 9.1.22) that have their correspondents on the counter-façade of the narthex (Figure 9.1.25), which on the contrary has lost most of the traces of holes and recessions because of the changes that occurred over the centuries. The usefulness of such smaller holes is not clear yet, but they were surely made for a specific purpose.

Above each of the greater holes it is possible to see another rectangular hole, approximately 28 cm wide, and about 3 m distant from the corresponding one at the lower level

Figure 9.1.26 Holes for rafter ends.

(Figure 9.1.26). This distance corresponds exactly to the height of the trusses of the aisles. Therefore, these holes, now filled with stones, hosted the upper end of the rafters of these half trusses.

It is worth highlighting that this type of roof was much lighter than the vaults built by the Crusaders.

The surveys carried out during the excavation campaign showed that the cross vaults are made of irregular and roughly cut stones, with a variable thickness between 35 and 40 cm, and are connected to the façades of both narthex and church only in correspondence of the corbels at the base of the vaults and for a stretch not greater than 1,10–1,30 m (Figure 9.1.27).

Instead, such vaults are well connected with each other in the north–south direction, by means of a system of main arches, orthogonal to both façades, and therefore in the west–east direction, and built with greater and more regular stones, as usual in similar local cross vaults. These stones have approximately the same dimensions, but they are slightly offset from each other so as to form wedges that allow a greater clamping with the stones of the vaults (Figures 9.28, 9.29). There are also other arches on the diagonal planes of the vaults; they are made with more regular stones, although not as the ones of the main arches (Figure 9.1.29).

In consequence of the rotation of the façade of the narthex, the cross vaults, in partic-ular the most central ones, lowered in the central zones, detached from the façade walls and cracked both at the extrados and the intrados (Figure 9.1.30). Especially the central vault exhibited detachments of 17–19 cm from the narthex façade and 10–11 cm from the church façade. Moreover, big cracks, parallel to the façades, were visible at the extrados

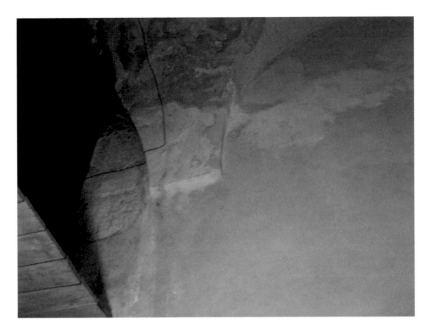

Figure 9.1.27 Corbel at the base of the vault.

Figure 9.1.28 Main arches in west–east direction.

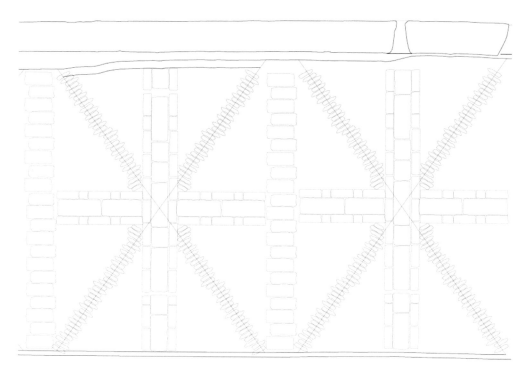

Figure 9.1.29 The main arches with offset stones and diagonal arches.

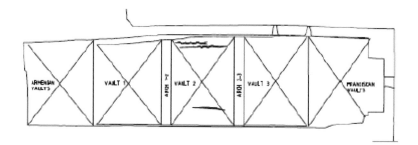

Figure 9.1.30 Crack pattern on the extrados of the vaults.

(Figures 9.1.30, 9.1.31). The detachment of the vaults from the walls was made easier by the lack of any connection between walls and vaults, the latter being added in later time and built in simple adherence. Therefore, it is possible to state that the extremely high out-of-plumb of the façade of the narthex is due also to the absence of those internal constraints that only a good connection of orthogonal walls and vaults with the façades could guarantee.

Even the arches connecting the vaults to each other in the east–west direction are strongly deformed. Moreover, the intrados of the vaults suffered water infiltrations because of these cracks and gaps and also for the absence of any maintenance at the extrados (Figure 9.1.12).

Figure 9.1.31 Cracks and detachments from both façades.

The causes of the rotation of the narthex façade towards the square were not clear. Therefore, it seemed appropriate, as a first approach to the problem, to check the possible dependence of such a rotation on a soil subsidence and on the state of the foundation. To this purpose, an inspection well was dug manually in the south-west corner of the surveillance office, just behind the buttress, in the area corresponding to the greatest rotation (Figure 9.1.32).

The floor tiles were carefully numbered before their removal (Figure 9.1.33). They rested on a thin sandy layer (about 2 cm), below which there was a lime-based mortar layer with low compactness and small thickness.

During the excavation a considerable amount of ceramic material, small mosaic tiles in white stone, wooden fragments and metal residues were found.

The excavation, carried out up to 1.85 m under the floor level (Figure 9.1.34), showed that the wall of the narthex façade continues downwards with a constant thickness up to a depth of 94 cm from the floor, beyond which there is a sort of foundation 47 cm high, made of bigger stones protruding about 20 cm. A further stone floor was found at the maximum depth, probably belonging to the first basilica. However, nothing was noticed that could justify the dependence of the rotation of the façade on the soil subsidence or on some failures in the foundations.

The causes of the damage that occurred to the façade of the narthex and, consequently, to the vaults remain still unknown. However, the replacement of a light wooden roof with

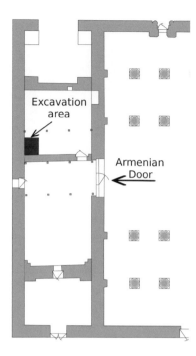

Figure 9.1.32 Excavation area inside the narthex.

Figure 9.1.33 Numbering of the floor tiles.

a heavy system of thrusting masonry vaults probably gave a significant contribution to the rotation of the façade towards Manger Square. It is believed, however, that such thrusts alone could not cause horizontal displacements as large as those measured. Because Bethlehem is in a seismic area, which over the centuries underwent several earthquakes, some of them even severe, it is very probable that an earthquake or a series of earthquakes, probably occurring from the 16th century onwards, were among the main causes of damage, as proved by recent numerical simulations [6].

Figure 9.1.34

Excavation up to the foundation level.

9.2 On-site material survey

C. Alessandri and M. Martinelli

During January 2015 an accurate campaign of on-site investigations was carried out by means of visual and instrumental surveys of geometric type (laser scanning and traditional surveys), of mechanical type (sonic, thermographic and hygrometric tests, tests on mortars), destructive tests in the laboratory and tests of classification of the masonry quality. This campaign concerned the analysis of the material properties of masonries and mortars in the walls and the vaults of the narthex and the control of the global state of conservation. The presence and the quality of connections among the various resistant elements were also verified in the context of the geometric survey. The collected data, most of which was indispensable to carry out structural analyses and numerical simulations, were added to those already reported in the official documents provided in the past (Restoration of the Roof of the Church of Nativity – Stage 2 Study and Assessment Report January 2011, Restoration of the Roof of the Church of Nativity – Final Report July 2011, Restoration of the Roof of the Church of Nativity –Appendix D "Scope of works, Drawings and Technical specification" April 2013). They provided a deep knowledge of the geometric and material properties of all the components of the narthex and turned out to be extremely useful to study the structural behavior of the whole narthex under static and dynamic loads and to design the necessary interventions of consolidation.

As for the walls, for example, it was possible to ascertain that the transversal walls delimiting the Armenian (south) bay and S. Helena chapel are not connected to the façades of the church and the narthex, which therefore are constrained only at their extremities in correspondence of the north and south sides. Even the closure of the lateral doors in both façades was made with poor-quality masonry, not effectively connected to the masonry of the walls, thus providing only a modest increase in stiffness, especially in the presence of horizontal actions orthogonal to the façades. The deformations of these façades have already been mentioned before. In the narthex façade in particular, to a careful observation, no cracks in the plaster were surveyed, which proves that the phenomenon was already over in the period prior to the last plastering made in the 19th century. Moreover, the external buttress built in the 18th century is perfectly vertical so demonstrating that no further rotation occurred after its construction. However, some inspections carried out where the narthex façade starts bending have shown some interventions in the mortar joints that testify to some repairs carried out before the last plastering.

The façade of the church is about 1.15 m thick and is made of squared stones. It is probably without internal filling or with very limited filling. The masonry texture of all the walls of the narthex was surveyed. In particular, the texture on the internal vertical surfaces was detected by removing entire portions of plaster up to the underlying stones. It is worth noting that this plaster, dating back to the 19th century and with no artistic or historic value, had been seriously damaged by moisture and rainwater infiltrations. The side walls, about 1.00 m thick, are made of roughly cut stones and have an inner core, as proved by the sonic tests. The façade of the narthex has a thickness of about 1.15 m and is made of squared stones, like those of the façade of the church. The buttress, on which no instrumental investigations was carried out because of its big dimensions, appears to be made of roughly cut stones with a probable wide, if not very large, internal core. The two orthogonal walls inside, probably added to support the two external bell towers, as already said in the previous chapter, are made of lower-quality masonry if compared to the remaining walls and are not clamped to the façade walls of narthex and church.

The same types of investigations were carried out in the vaults and new data were obtained in addition to the ones already mentioned previously. The parts of the vaults enclosed between the diagonal arches are made of more irregular and generally smaller stones with a greater percentage of mortar. The vault 1 (the one close to the Armenian bay) seems to be built with a lower-quality masonry and with a different technique. Because of the lack of the diagonal arches and of visible keystones at the extrados it seems that it was partially rebuilt. In some parts the vaulted shape is even barely recognizable. The sonic tests carried out on the vaults have provided extremely low speed values, which means the presence of incoherent masonry with a great prevalence of voids and/or very poor-quality mortar.

Inside the narthex, the removal of the plaster in correspondence of the corbels (the lower part of the vaults and arches) of vaults 1 and 2 allowed verifying that they are made with well-worked and big-sized stones embedded in the walls that had been appropriately worked to host them. A corbel on the narthex façade exhibits some vertical cracks due to crushing, which however were repaired with mortar before the plastering. Moreover, the absence of cracks on the plaster proves that the structure was stable at the time of the survey. The corbels on the church side present no problems and appear perfectly intact. The analysis of the intrados of these vaults was more difficult, if not impossible, due to the presence of the existing propping in the central bay and of those added in the side bays at the beginning of the survey campaign in order to guarantee safe work conditions. Moreover, the urgency to prop vaults 1 and 3 in order to proceed quickly with the survey of the extrados did not allow to carry out in advance an extensive investigation on the entire intrados of the vaults. The control of the state of conservation of the intrados was therefore carried out on the basis of photos and local inspections by removing some wooden boards covering the intrados and portions of plaster. The data collected allowed stating that there were no significant cracks on the intrados.

According to the Italian Standards [7], which were taken as a main reference for the technical checks, visual and instrumental analyses should have allowed the adoption of a knowledge level LC2 for the determination of the mechanical characteristics of the materials. However, not all the in situ tests required by the Standards could be performed. In particular, it was not possible to carry out semi-destructive tests with flat jacks to determine the mechanical characteristics of masonry, both for the difficulty in finding the necessary equipment in the area and for the designers' intention not to further alter the already precarious conditions of the masonry. In place of these, the values of the mechanical characteristics were derived from a series of visual investigations that led to the definition of the so-called Masonry Quality Index (IQM)* [8]. Consequently, the LC1 knowledge, a more precautionary knowledge level, was chosen for the benefit of safety. Moreover, Fc = 1.35 was assumed as the correspondent confidence factor. This assumption also takes into account the fact that the characterization of the local masonry was made with reference to technical tables and evaluation criteria defined for Italian architectural contexts, which are quite different from the local ones.

IQM

The Masonry Quality Index is simply a number by which the degree of compliance of a masonry sample is assessed to what is expected from a correct execution according to the so-called rule of art, i.e. the set of constructive measures and expedients that guarantee the good behavior of a masonry and ensure its compactness. It is defined on the basis of some parameters surveyed in the examined sample, such as the quality of mortar, the presence of lithic elements in the orthogonal direction to the wall, the shape and size of the stones, the greater or minor alignment of the vertical joints, the presence of regular horizontal rows of mortar joints, the strength of the components. The IQM is evaluated separately for three possible different types of actions: vertical loads, horizontal actions in the plane of the wall and horizontal actions orthogonal to the wall. The parameters listed above differently influence the response of the wall to the different types of action. For this reason, three different IQM values are defined: vertical IQM, IQM in the plane of the wall and IQM orthogonal to the wall. For each wall sample, in correspondence of each IQM value, it is possible to define, by using appropriate reference tables, the values of the average compressive strength f_m, of the average shear strength τ_0 and the average value of the longitudinal module of elasticity E, all parameters that can be used in the numerical verifications.

9.2.1   Sonic tests

The sonic tests are non-destructive investigations capable of characterizing and qualitatively describing the masonry. They are based on some relationships that bind the propagation velocity of elastic waves in a medium to the elastic properties of the medium itself.

The propagation velocity of the elastic waves is directly related to the mechanical and physical characteristics of the material if this is elastic, homogeneous and isotropic. The elastic waves used for sonic investigations on masonry have frequencies between 20 Hz and 20000 Hz and are generated by a transducer (hammer) that produces vibrations in the impact with the material to be tested. These waves are detected by a sensor, located on the opposite side of the sample, at the same height as the transmitter, so that both devices are along the propagation direction of the waves. The received pulses are then displayed and recorded by an acquisition device (handheld computer with oscilloscope) (Figure 9.2.35).

The part of the masonry to be analyzed is previously freed of any plaster, and the points to be beaten by the hammer are reported on both sides according to a grid with predefined mesh. The signal captured by the receiver contains all the information related to obstacles or discontinuities encountered, such as zones with different degrees of aggregation, voids, areas with material density reduced by degradation phenomena. In all these cases there is a decrease in the propagation speed of the elastic wave generated by the impact of the hammer on the masonry.

The measurement campaign was carried out by using the Boviar CMS v.3 sonic/ultrasound acquisition system, consisting of a data acquisition unit and piezoelectric sensors with instrumented hammer, in order to carry out measurements of the propagation time of compressive waves (P waves) in many types of materials, even with poor propagation and speed characteristics. All the displayed signals were memorized directly on an HP Ipaq Handheld, on which the Datasonic v.2.5.5 software, which manages display, storage and processing of the signals, had been loaded. All the tests were performed for transparency by using geometric grids of variable numbers of points, from a minimum of 24 points to a maximum of 78 points. The mesh size was 15 cm. On one side of the wall the points were drawn by using a grid previously drawn on a cardboard; the corresponding points on the opposite side were defined by using a magnetic transmitter/receiver system (HILTI PX 10) (Figure 9.2.36).

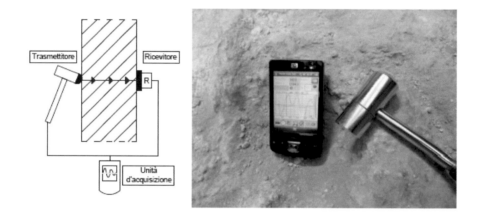

Figure 9.2.35 Hammer, sensor-receiver, acquisition device.

Figure 9.2.36 Grid tracking with cardboard and transmitter/receiver.

9.2.2   Hygrometric test

The method to measure moisture levels exploits the phenomenon of propagation of electro-magnetic waves in a medium. The measurement is based on the principle of reflection. The instrument emits an electromagnetic wave that invests a volume of material that can be com-pared to a cylinder 20–30 cm high and with a diameter of about 10–15 cm (Figure 9.2.37). The device records the characteristics of both the emitted and the reflected waves, providing a numerical index linked to the moisture content of the wave reflected by the medium. The test points are arranged on a grid with a mesh of 25 cm, such as to contain a cylinder of the diameter defined before. In this way there is no interference between the material volumes measured at two consecutive points, and consequently no influence of the measurement at one point on the data detected at other points. A minimum lateral distance of 10–15 cm with respect to the edges of the wall to be checked is also considered to avoid interference with adjacent structural elements with different thickness.

In the narthex sonic and hygrometric tests were carried out in the same areas where the visual inspections for the determination of the IQM had been carried out, so as to provide as much information as possible about the internal structure and the quality of the masonry.

In particular, the IQM was determined in three parts inside the narthex, in the façade wall of the church (IQM 1), belonging to the original body of the church, in the internal wall separating the south bay from the rest of the narthex (IQM 2), and in cross vault n. 1 (IQM 3; Figure 9.2.38). The sonic tests (NC N (Nativity church – narthex): SO1, SO2, SO3, SO4, SO5, SO6, SO7, SO8, SO9, SO10) were carried out in a greater number of areas, including those analyzed for the IQM (Figure 9.2.39). Of the hygrometric tests UMID 1, UMID 2 and UMID 3, only UMID 1 and UMID 2 were carried out in areas analyzed for the IQM. In the sequel the results of the sonic tests will be shown for the same areas analyzed for the IQM in order to show how both tests can provide more exhaustive and reliable information on the material properties.

IQM 1

The wall of the façade of the church is made of big, regular stones, the so-called malaki, a local lime stones extracted from quarries in the territory of Bethlehem, with appropriately staggered mortar joints perfectly adhering to the stones, although with some tendency to pulverise for the presence of salts and moisture (Figure 9.2.40).

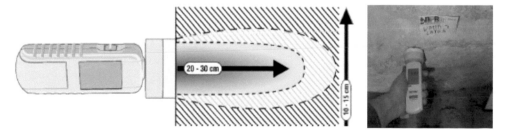

Figure 9.2.37 Transmitter/receiver device ad volume invested by the electromagnetic waves.

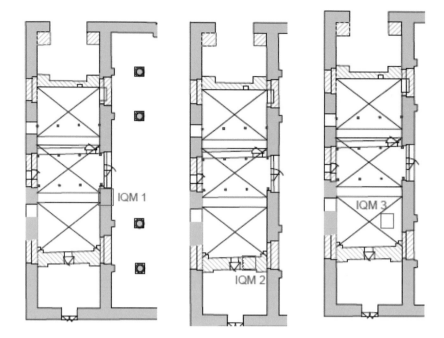

Figure 9.2.38 Samples for IQM evaluation.

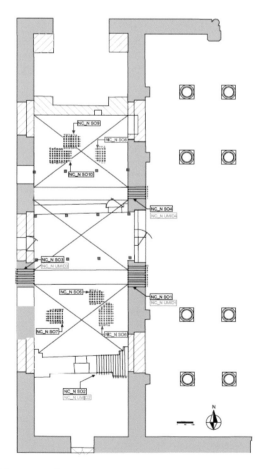

Figure 9.2.39 Sonic and hygrometric tests.

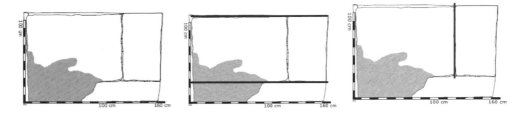

Figure 9.2.40 Sample for IQMI.

Table 9.2.1 IQM I values for vertical, out-of-plane, in-plane actions

VALORI STIMATI PER IQM PARI A 8,5 – AZIONI VERTICALI

Param. meccanici	MIN	MED	MAX
f_m (N/cm²)	629	774	919
E (N/mm²)	2411	2858	3305
τ_o (N/cm²)	12	14,28	16,55
G (N/mm²)	778	922	1065

VALORI STIMATI PER IQM PARI A 7,22 – AZIONI FUORI DAL PIANO

Param. meccanici	MIN	MED	MAX
f_m (N/cm²)	471	592	712
E (N/mm²)	1928	2303	2679
τ_o (N/cm²)	9,06	10,93	12,8
G (N/mm²)	622	743	864

VALORI STIMATI PER IQM PARI A 7,65 – AZIONI COMPLANARI

Param. meccanici	MIN	MED	MAX
f_m (N/cm²)	520	648	776
E (N/mm²)	2080	2477	2875
τ_o (N/cm²)	9,97	11,96	13,96
G (N/mm²)	671	799	927

Table 9.2.1 shows the minimum, mean and maximum values of the compressive strength f_m, Young modulus E, shear strength τ_0 and shear modulus of elasticity G for each IQM value related to vertical actions, out-of-plane actions and in-plane actions.

The sonic tests (SO1) carried out in the same masonry sample denoted the presence of stones placed orthogonally to the wall, in particular in the right part of the sample (Figure 9.2.41) where the sonic velocities achieve the highest values. As a matter of fact, this part is nearly adjacent to a corbel where there is a greater concentration of pressures. In the remaining part of the sample, where the velocities are much lower, the masonry seems to be more disconnected inside and probably made with a double curtain of stones and an internal core of filling material.

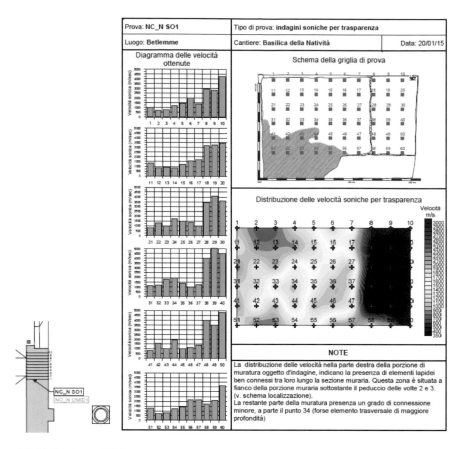

Figure 9.2.41 Sonic test SO1 – output.

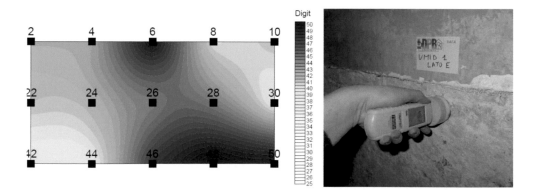

Figure 9.2.42 Hygrometric test UMID 1 – output.

The measurements made in the same area with a contact hygrometer (UMID 1) showed no significant variations in the speed at which the impulses passed through the medium, such as to justify moisture concentrations (Figure 9.2.42).

IQM 2

The wall analysed for IQM 2 is made of big stones, although not so regular as those in the façade of the church (Figure 9.2.43). Also, the horizontal joints are not always on the same plane for the whole length of the sample, and that can justify, along with their variable thickness, the hypothesis of a wall built and modified in different times. Even the distance between one row of horizontal joints and the other changes considerably, whereas the vertical joints are distributed with some irregularity depending on the different sizes of the stone.

Table 9.2.2 shows the minimum, mean and maximum values of the compressive strength f_m, Young modulus E, shear strength τ_0 and shear modulus of elasticity G for each IQM value related to vertical actions, out-of-plane actions and in-plane actions.

The sonic tests SO2 (Figure 9.2.44) in this area denote a poor connection among the stones within the thickness of the wall and therefore the possible presence of a double curtain of stones with a weak internal core and connected from time to time by stones in the orthogonal direction. In particular, the masonry shows a worse internal connection in the right part of the sample where the velocities decrease considerably. It is worth noting that this area is quite delicate because it is located under the base of the vault where the level

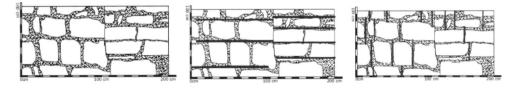

Figure 9.2.43 Sample for IQM 2.

Table 9.2.2 IQM 2 values for vertical, out-of-plane, in-plane actions

VALORI STIMATI PER IQM PARI A 7,22 – AZIONI VERTICALI

Param. meccanici	MIN	MED	MAX
f_m (N/cm²)	471	592	712
E (N/mm²)	1928	2303	2679
τ_0(N/cm²)	9,06	10,93	12,8
G (N/mm²)	622	743	864

VALORI STIMATI PER IQM PARI A 5,53 – AZIONI FUORI DAL PIANO

Param. meccanici	MIN	MED	MAX
f_m (N/cm²)	326	417	509
E (N/mm²)	1439	1736	2034
τ_0(N/cm²)	6,3	7,72	9,15
G (N/mm²)	464	560	656

VALORI STIMATI PER IQM PARI A 6 – AZIONI COMPLANARI

Param. meccanici	MIN	MED	MAX
f_m (N/cm²)	358	457	556
E (N/mm²)	1556	1872	2189
τ_0(N/cm²)	6,9	8,45	10
G (N/mm²)	502	604	706

of the compressive stresses is higher. Therefore, here an intervention of consolidation was planned in order to guarantee a higher value of the material strength.

As for the measurements made in this area with a contact hygrometer (UMID 2), a narrow band was found (points 3–29–55) with a higher moisture content than in the surrounding area (Figure 9.2.45), but not such as to influence the results of the sonic investigations.

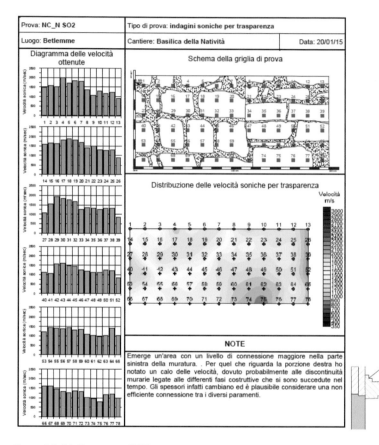

Figure 9.2.44 Sonic test SO2 – output.

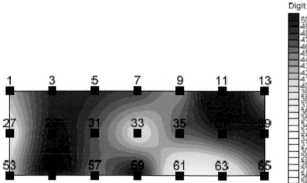

Figure 9.2.45 Hygrometric test UMID 2 – output.

IQM 3

The masonry of the vault is rather chaotic, with stones of various sizes and irregular shape. The smaller stones are present mainly inside the vault, and larger and better carved stones are inserted among them as wedges in order to keep them more tightened and increase the compactness of the whole vault. Despite this constructive shrewdness, the level of internal connection is quite low. Because of the variety of the stones in terms of size and shape the mortar joints are very thick and do not follow any regular pattern (Figure 9.2.46).

Table 9.2.3 shows the minimum, mean and maximum values of the compressive strength f_m, Young modulus E, shear strength τ_0 and shear modulus of elasticity G for each IQM value related to vertical actions, out-of-plane actions and in-plane actions.

The low sonic velocities surveyed in most of the area during the sonic tests (SO5) confirm the poor internal connection of the masonry (Figure 9.2.47).

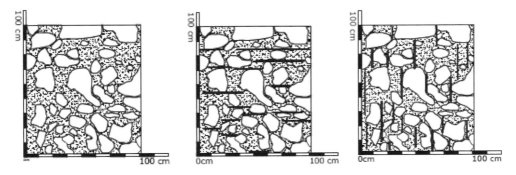

Figure 9.2.46 Sample for IQM 3.

Table 9.2.3 IQM 3 values for vertical, out-of-plane, in-plane actions

VALORI STIMATI PER IQMPARI A 1,7 – FUORI DAL PIANO

Param. meccanici	MIN	MED	MAX
$f_m(N/cm^2)$	137	187	238
E (N/mm^2)	739	913	1088
$\tau_0(N/cm^2)$	2,72	3,49	4,26
G (N/mm^2)	238	295	351

VALORI STIMATI PER IQM PARI A 2,12 – AZIONI NEL PIANO DELLA SUPERFICIE

Param. meccanici	MIN	MED	MAX
f_m (N/cm^2)	150	204	258
E (N/mm^2)	794	978	1163
$\tau_0(N/cm^2)$	2,98	3,8	4,62
G (N/mm^2)	256	315	375

VALORI STIMATI PER IQM PARI A 2,12 – AZIONI NEL PIANO DELLO SPESSORE

Param. meccanici	MIN	MED	MAX
f_m (N/cm^2)	150	204	258
E (N/mm^2)	794	978	1163
$\tau_0(N/cm^2)$	2,98	3,8	4,62
G (N/mm^2)	256	315	375

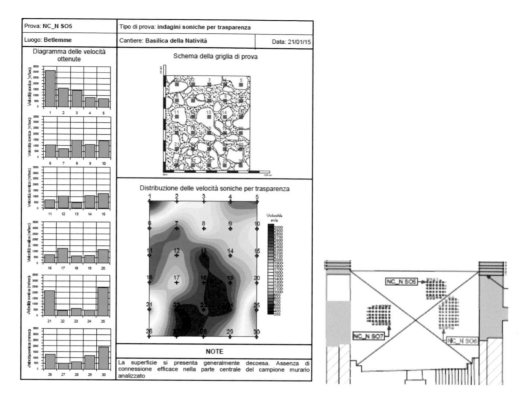

Figure 9.2.47 Sonic test SO5 – output

9.2.3 Thermography

Thermography allows us to carry out non-destructive analyses by using the capacity of each material to emit energy through electromagnetic radiations. Each element of a building, once heated, is subject to different thermal gradients or thermal variations according to the thermal conductivity and the specific heat of those parts that make it up. Based on the detected response it is possible to have information on some features of the analyzed element, such as its internal composition, the presence of internal cavities or moisture.

Thermographic analysis can be conducted passively, by analyzing natural thermal cycles (insolation and subsequent cooling), and actively, by artificially heating the surface to be observed. The test is carried out by recording a thermal image of the object through color or grey scales by means of a special device, called thermal camera, which is sensitive to infrared radiations. There is no physical contact with the analyzed surface. Each color or tone of the grey scale corresponds to a temperature range that is generally in the order of fractions of degree centigrade. In the specific case, a Trotec IC080LV model thermal imaging camera was used (Figure 9.2.48). To heat the surfaces, two LPG convectors were used at the same time. The environmental parameters of moisture and temperature were detected with ALMEMO MA2470 professional data-logger and FHAD46–2L00 digital thermo-hygrometric probe (Figure 9.2.49).

Figure 9.2.48 Thermal camera.

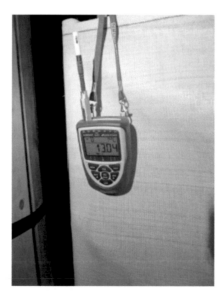

Figure 9.2.49 Digital thermo-hygrometric probe.

The investigation was carried out on some internal walls of the narthex, in particular on the three walls below the south vault (vault n. 1; Figure 9.2.50 a, b, c) and on the internal surface of the façade of the church (Figure 9.2.50 d). All these walls were analyzed for all their full height.

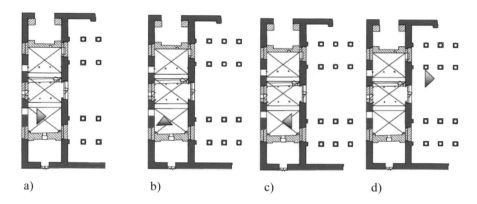

a)   b)   c)   d)

Figure 9.2.50 a, b, c walls under the south vault (vault n. 1) – d internal façade of the church.

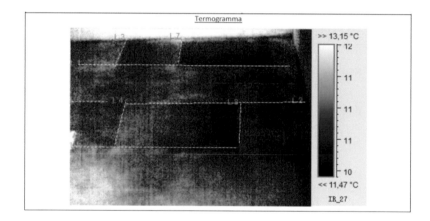

Figure 9.2.51 Masonry texture.

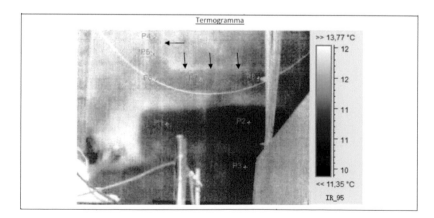

Figure 9.2.52 Infills.

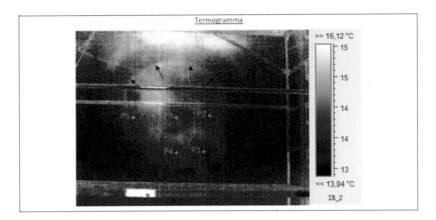

Figure 9.2.53 Relieving arches.

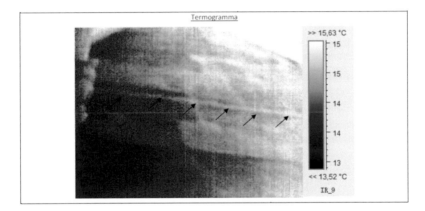

Figure 9.2.54 Wooden lacing.

The thermographic analysis made it possible to highlight plaster detachments from the wall support, variations in plaster thickness, masonry textures and irregularities (Figure 9.2.51), infills (Figure 9.2.52), relieving arches above the buffered lateral doors (Figure 9.2.53), wooden lacings in the wall (Figure 9.2.54).

9.2.4 Tests on mortar samples

Simultaneously with the tests described above, petrographic analyses were also performed on bedding mortars and plasters. These tests were carried out to determine with greater precision the quality of masonry in the vaults and in the walls of the narthex. Firstly, there was a visual analysis of the samples without using any instrument. The samples were then analyzed with the stereomicroscope in glossy section and with a polarizing microscope in thin section. These two techniques allowed drawing qualitative conclusions on the properties of the materials observed.

In this phase, the following instruments were used:

- a stereoscopic microscope with low magnification (commonly called stereomicroscope) Optika SZM-LED2, with LED lighting for observation in reflected light;
- an Optika B-600 POL polarizing microscope for transmitted light observation;
- an Optika Vision Pro camera;
- the Optika Vision Pro Plus 5.0 software.

The withdrawal areas were identified by taking into account their location, the absence of visible degradation, the simultaneity with sonic and thermographic analyses, the possibility to withdraw a suitable amount of material.

Each sample was coded with a progressive number and with reference to its location. i.e.:

> NC_N_V1–1:   Nativity Church, narthex, Vault n. 1, sample n. 1
> NC_N_M1:   Nativity Church, narthex, Mortar Sample n. 1

The withdrawal points for the samples are shown in the following plan (Figure 9.2.55).

The samples were collected manually, by using the hammer and chisel/awl, after carefully cleaning the surface to allow the removal only of the material to analyze. The first, summary impressions of the conditions of the fragments were noted on-site and later verified during the analysis in the laboratory with the microscope. The selected samples were dried and impregnated with epoxy resin in order to be prepared in a glossy section and in a thin section. For each sample a specific analysis card was prepared, showing the place of collection, the images of the sample before and after the withdrawal, the photographic documentation of the sample as it was and, where required, the analyses carried out. Of the 13 samples collected, 10 were analyzed in a glossy section with the stereomicroscope and in a thin section with a polarizing microscope. These samples were selected because they were considered representative of the various types of mortar/plaster present in the narthex.

These tests were carried out by the TS Lab & Geo-services studio in Cascina, Pisa.

Based on what was observed with the stereomicroscope and, in thin section, with the polarizing microscope, some conclusions could be drawn.

All the samples were made with aerial lime, whose nature remains uncertain, except for the NC_NV2–5 sample. To improve the characteristics of the mortars, some additives were added to the lime, like *cocciopesto* powder, the presence of which was surveyed in various fragments. The quality of the binder is however very poor. Its fragility is also aggravated by the scarce presence of aggregates, which led to the occurrence of shrinkage cracks, frequent in the samples analyzed and such as to weaken the mechanical characteristics of the mortars.

The aggregates used have the same nature, regardless of the type of mortar. They are made with limestone and come probably from the processing of the stones for the vaults and the walls.

In NC_N_M4, taken from the intrados of vault n. 1, cavities of elongated shape were found. They are probably due to the presence in the mixture of organic fibres (probably straw), now disappeared and no longer easily identifiable. Organic materials (such as straw,

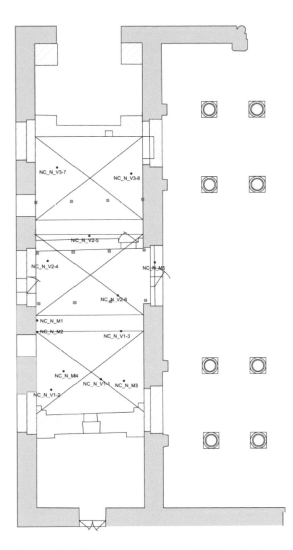

Figure 9.2.55 Withdrawal points: in RED on the extrados, in BLUE on the intrados.

in fact) were added to the mixture in the past to prevent shrinkage cracks. However, this type of cavities was not found in any of the samples analyzed.

In conclusion, the mortars used for the vaults (samples NC_N_V1–1, NC_N_V1–2, NC_N_V2–5, NC_N_V2–6, NC_N_V3–7, NC_N_V3–8, NC_N_M4) are of low quality, little cohesion and poorly adherent to the stone elements. The same conclusions can be drawn for the sample NC_N_M1, a bedding mortar taken in the area where the sonic test SO3 was performed. The NC_N_M3 fragment is the only one that shows two layers: the first, more internal, very weak and characterized by wide cracks; the second, external, very rigid and made in recent periods with artificially prepared aggregates. None of the materials observed shows alteration products. That allows to suppose that the climatic conditions in which the mortars have been maintained have not undergone substantial changes over time or that there have been no external agents such as to cause a profound deterioration of the mixture components.

9.2.5   Tests on stone cores

Two cores with a 10 cm diameter were taken from the vaults. Both specimens had the same lithology and textural characters and could be defined as vacuolar and laminated travertine limestone. No sulphate components, like gypsum, nor amorphous silica or phyllosilicates, were found. Each sample as a whole was consistent and not altered, although locally they were not very cemented.

Moreover, other 4 stone cores were taken from the vaults to carry out uniaxial compression and tensile tests. From one core two cylindrical specimens were obtained, one for the uniaxial compression test and the other for the tensile strength test (Brazilian test). From each of the remaining three cores one cylindrical specimen was obtained for the uniaxial compression tests. All the specimens had diameter and height of 1 cm.

Only the values of the ultimate strength in tension and compression are reported here.

Ultimate strength in tension: 0.64 N/mm^2
Ultimate strengths in compression (N/mm^2): 8.7; 5.8; 25.7; 14.4

These results, together with those obtained from the investigations previously presented, allowed defining quantitatively the mechanical parameters of masonry in the different parts of the narthex. On the basis of these parameters, the designers were able to more accurately define the interventions and the consolidation techniques and to carry out with a high degree of reliability all the structural checks required by law.

9.3   Interventions of restoration inside the narthex

C. Alessandri and M. Martinelli

At the beginning of the restoration works, the internal surfaces of the narthex in the bays corresponding to vaults 1 and 2 were covered by a plaster dating back to the 19th century. On the contrary, the surfaces of the north and south bays had no plaster and allowed seeing the whole masonry apparatus. The existing plaster showed in many parts detachment from the support, lacunae, efflorescence and moisture spots due to the infiltrations from the vaults. Given the beauty of the masonry apparatus in the surfaces that were still uncovered and imagining the large amount of information that the existing plaster might still hide, it was decided to remove this plaster from all the walls without replacing it with another plaster. Only the plaster on the intrados of the vaults, rather damaged and deteriorated, was to be replaced, also to hide the intrados stones, which were roughly cut and assembled in a chaotic way, and therefore without any aesthetic value. The new plaster would have been also used to hide some traces of the consolidation intervention that was under study.

The removal of the plaster from the walls was carried out by following the standard procedures and taking care not to damage the underlying surfaces in any way. Then the stones, mainly calcareous blocks of different sizes, were cleaned by removing incoherent deposits

Figure 9.3.56 Buffered side window – narthex counter-façade.

with flat brushes and small extractor fans and water. The consolidation intervention consisted of filling cracks and voids left by the loss of the original mortar with a new mortar compatible for color and grain size with the original one. Where it was necessary to recover the perception of the original state for a better reading and interpretation of the architectural elements, an integration of some missing pieces of stone was made with material compatible with the whole context and easily distinguishable.

The removal of the plaster made it possible to admire again, from inside, the magnificent wall structure of Justinian's time and the changes made over time. In particular, to the left of the Door of Humility (south-west side) it is possible to notice the buffering of one of the two windows that flanked the main entrance (Figure 9.3.56) as well as the buffering of one of the two side doors that led into the atrium, now partially hidden by a half pilaster added later (Figure 9.3.57).

On the opposite side (south east side) there is the corresponding side door, also buffered, that led into the nave (Figure 9.3.58).

On the south side it is now possible to see the masonry apparatus of the wall separating the central part of the narthex from the Armenian bay and built in the Crusader time (Figure 9.3.59).

It is interesting to note the presence of a relieving arch above the lintel of the main doors. The space between the arch and the lintel is filled with stones. Analogous arches could be surveyed also above the wooden lintels of nave and aisles (Figure 9.3.60). In this case their presence had already been detected by Hamilton and reported in his treatise [3].

Figure 9.3.57 Buffered side door – narthex counter-façade.

Figure 9.3.58 Buffered side door – church façade.

Figure 9.3.59 Crusader wall delimiting the Armenian bay.

Figure 9.3.60 Relieving arch above a wooden lintel.

Figure 9.3.61 Back of the main portal with Justinian lintel, Crusader arch and Humility Door.

Figure 9.3.62 Intrados of the lintel (flat arch).

At the centre of the counter-façade of the narthex it is possible to see the back of the main entrance with the lintel of Justinian's portal, in the form of a flat arch (Figure 9.3.61), of which it is now possible to admire the extraordinary cut of the perfectly squared stones through a void left in past ages and hidden by the plaster (Figure 9.3.62). Under the lintel there is the arch of the Crusader door and in the end the small Humility Door.

Close to the Humility Door and the Armenian door a wall built in the last century separates the main body of the narthex from the bay corresponding to vault n.3 and now occupied by the Surveillance Personnel (Figure 9.3.63).

Beyond this wall, in the bay corresponding to vault n.3 and unfortunately no longer accessible, it is possible to observe the second side door that introduced into the left aisle and the size of which was reduced in a period not yet defined (Figure 9.3.64). Above, however, it is possible to notice a mighty lintel surmounted by the usual relieving arch (Figure 9.3.65).

Figure 9.3.63 Wall between the guardroom and the rest of the narthex.

Figure 9.3.64 Internal side door.

Figure 9.3.65 Relieving arch above the lintel.

Figure 9.3.66 External side door.

On the opposite side, again in the counter-façade of the narthex, there is the external side door corresponding to the internal door mentioned above. Even this door was reduced in size before being filled completely with roughly cut and differently sized stones (Figure 9.3.66).

An arch of the Crusader time probably connected this bay with St. Helena's Chapel in the north bay of the narthex (Figure 9.3.67).

It is worth noting that, unlike the central parts of the narthex, the walls of this bay had never been plastered inside. Therefore, it was possible to start immediately with the interventions of cleaning and consolidation of stones and mortars. Moreover, the excavation made in this area to survey the foundations of the narthex façade allowed discovering part of the floor mosaics of Constantine's church (Figure 9.3.68). For all this valuable information that

Figure 9.3.67 Crusader arch between St. Helena's Chapel and the guardroom.

Figure 9.3.68 Fragment of Constantine's floor mosaics.

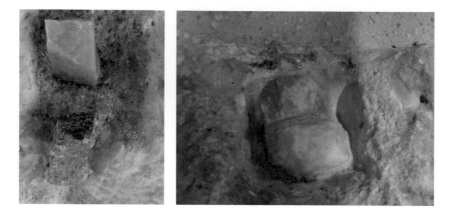

Figure 9.3.69 Marble and metallic elements for marble slab anchorage.

this space contains and that should be everyone's heritage it is desirable that the competent authorities find sooner or later the most appropriate way to make it finally accessible.

During the removal of the plaster a lot of coupled marble pieces and metallic elements were found under the plaster and still fixed to the walls at regular intervals (Figure 9.3.69). They formed the anchorage system to the walls of the marble slabs that covered the internal surfaces of the walls up to the intrados of the vaults. They were found in great quantity also in the walls of the aisles that, therefore, were originally all covered with marble.

9.4   Interventions of structural consolidation

C. Alessandri and M. Martinelli

The consolidation works started on June 19, 2015. The designers and the Presidential Committee had already decided to preserve the Crusader vaults for their historical meaning and for being after so many centuries one of the most characterizing elements of the narthex. It was also already known that the replacement of the original wooden roof with the masonry cross vaults, much heavier and, first of all, thrusting against the walls, had given an undoubted contribution to the rotation of the narthex façade, especially when the thrust of the vaults turned out to be associated with the horizontal seismic actions. The vaults, as already said before, had been seriously damaged, with cracks and detachments particularly visible in the central area, and all materials, mortars in particular, had undergone a considerable decay over centuries. The main idea behind the whole structural project was to separate the vaults from the floor of the terrace in order to let the vaults bear only their own weight. The self-weight of the floor and other additional vertical loads would be borne by an independent steel structure and transferred through it to the walls along the perimeter.

The restoration project of the narthex also envisages a significant increment of its resistance against seismic actions for Palestine being an area subject to seismic risk. This objective is achieved through the new steel structure mentioned before, connecting the façade of the narthex with the one of the church, and also through the reduction of the seismic masses represented by the filling incoherent material located at the extrados of the vaults and acting as a foundation to the paving of the terrace. This choice, while it is certainly beneficial in the presence of seismic actions, can be counterproductive for static actions, since the filling, as is known, by reloading with its own weight the flanks of the vaults, has a stabilizing effect on the masonry walls. For this reason, numerical checks were carried out on the vaults with reference to load conditions before and after the removal of the filling.

After removing from the extrados most of the filling material, the preliminary step before any further intervention was consolidating the masonry vaults. They were reinforced at the extrados by using a consolidating mortar injected through predrilled holes located according to a predefined square mesh (Figures 9.4.70, 9.4.71).

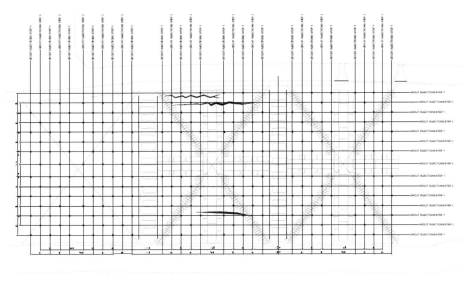

Figure 9.4.70 Square mesh for mortar injections.

Figure 9.4.71 Predrilled holes for mortar injections.

Figure 9.4.72 Glass fibre net and grout layer on the extrados of the vaults.

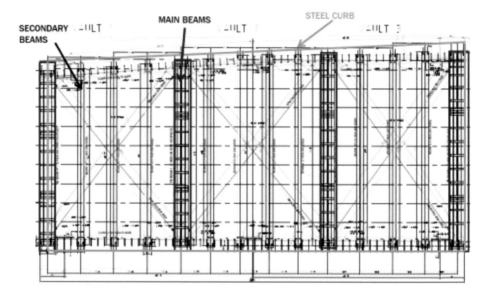

Figure 9.4.73 The new steel structure.

Then, the extrados was coated by a grout layer and a net of glass fibers (Figure 9.4.72).

The new steel structure is placed above the three central bays and is made of four main beams (above the stone arches placed between one vault and another) and 10 secondary beams parallel to the main ones (Figure 9.4.73). As said before, it bears the weight of the terrace floor and is properly anchored to the walls so as to act as a reinforcement of the whole masonry structure against horizontal loads generated by earthquakes.

All the beams rest on the two opposite façades with their ends inserted in proper spaces already existing or obtained by removing or cutting some stones (Figure 9.4.74). The ends of the beams are connected also to a steel curb fixed to the perimeter of the space corresponding to the three central bays (Figure 9.4.75).

Figure 9.4.74 Main and secondary beams resting on both façades.

Figure 9.4.75 Steel curb along the whole perimeter.

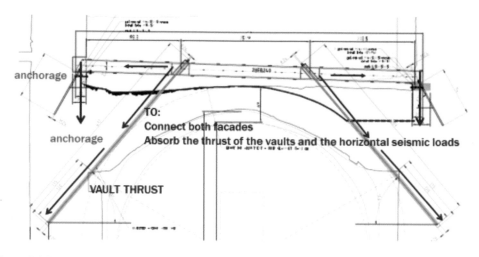

Figure 9.4.76 Extrados tie rod.

Moreover, all the beams are strongly connected to the façades by means of pairs of short inclined bars placed at the ends. In the main steel beams there are also pairs of inclined steel rods coming out from about one-third of the length of the beams and such as to lap the main arches between one vault and another (Figure 9.4.76). The system "horizontal beam–inclined steel rods" is the well-known extrados tie rod which is entrusted with the task of containing the static thrust, however small, exerted by the vaults and the thrust produced by seismic actions. This type of extrados tie rod, although less efficient than the classic intrados tie rod, has been widely used in the past to avoid the encumbrance and visibility of the latter. The high stiffness of the beams to which the diagonal tie rods are anchored reduces vertical displacements to the advantage of the efficiency of the extrados tie rod. The connections between the two façades, provided by the steel curb and the anchorage with inclined bars, minimize the differential horizontal displacements occurring in the E–W direction in the presence of an earthquake, thus reducing considerably the risk of collapse of the whole structure.

The vaults were also connected to the intrados of the beams by means of vertical steel tie rods (Figure 9.4.77) inserted from the extrados in predrilled holes according to a defined orthogonal grid (Figures 9.4.78, 9.4.79).

These thin rods (Figure 9.4.80) form a further safety measure in addition to the actual seismic reinforcements and come into action only in the occurrence of some incipient collapse due to some unexpected and uncontrolled special events. Moreover, the uncertainties related to the actual state of conservation, strength and stiffness of that masonry before and after the consolidation were too high not to require an additional device for greater safety.

In the end, the reinforcement of the intrados was completed, as for the extrados, by means of a grout layer and overlapping pieces of glass fiber net in order to form a protective coating incorporating the ends of the vertical bars (Figure 9.4.81).

The works ended in April 2016 with the installation above the steel structure of the corrugated steel sheets (Figure 9.4.82), light concrete, steel net, waterproofing layer, sand layer and tiles (Figure 9.4.83).

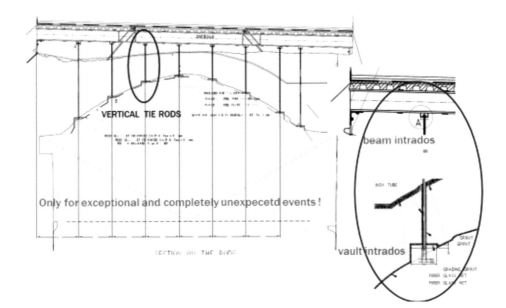

Figure 9.4.77 Vertical steel tie rods.

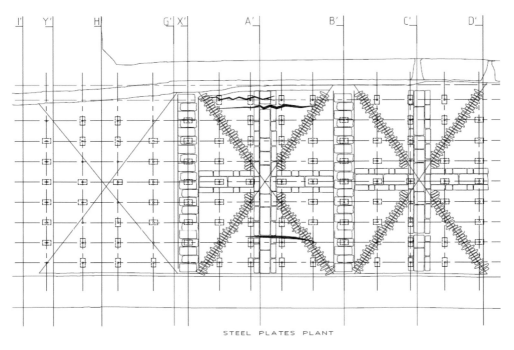

Figure 9.4.78 Orthogonal grid for vertical tie rods.

Figure 9.4.79 Drilling for tie rod insertion.

Figure 9.4.80 Tie rods details.

Figure 9.4.81 Intrados grout layer and glass fiber net.

Figure 9.4.82 Corrugated steel sheets.

Figure 9.4.83 Final layer of tiles.

Figure 9.4.84 Inspection wells.

Some inspection wells were provided in correspondence of the cable/rod tensioners and for general maintenance of the steel structure under the floor (Figure 9.4.84).

9.5   Numerical simulations and on-site tests

C. Alessandri and M. Martinelli

9.5.1   Vaults and walls

Even before the acquisition on-site of the material data, some comparative analyses were carried out on the vaults before the intervention (deformed vaults with the existing filling on flanks) and the same vaults after the intervention (deformed vaults with reduced filling) to evaluate for both cases the safety levels of the structure in the presence of vertical loads. These analyses, preliminary and preparatory to more refined modelling, were performed by using the software "ARCO" [9] based on J. Heyman's theory [10], readjusted and developed for use on personal computers. As known, they are based on hypotheses of equilibrium and do not require the knowledge of the mechanical characteristics of masonry. In order to prevent the occurrence of hinges, checks were made on the thrust line position inside the thickness of the arch and on the compressed section percentage. The software provided also the highest values of the compressive stresses to compare to the strength values of the masonry. It is worth noting that this software solves only plane structures, not three-dimensional structures. Therefore, each vault quarter was divided into an appropriate number of arches, each subject to its own load condition. Other arches analyzed were the diagonal arches of the vaults and the arches connecting two adjacent vaults. The numerical results confirmed the stability of the vaults in both load conditions (with the existing filling and with reduced filling). However, even with the design load arch 2-3   min % compressed section = 36.7%> 25%   safety coefficient = 1,468

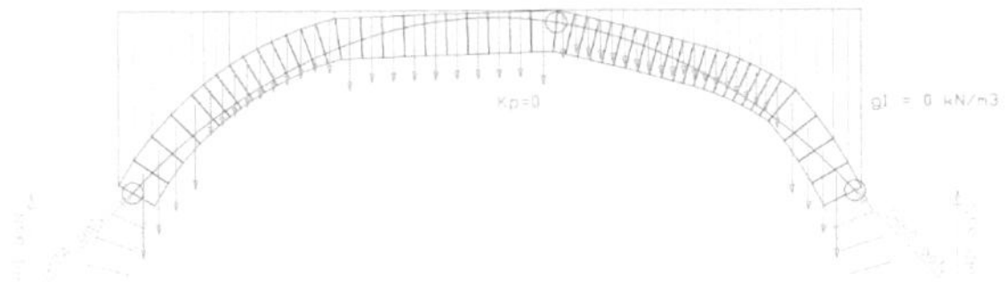

Worst values		Sec.N*
σ e [MPa]	0.401	22
σ i [MPa]	0.401	45

Figure 9.5.85 Arch 2–3 – thrust line.

diagonal arch 1   min % compressed section = 29.6%> 25%   safety coefficient = 1,184

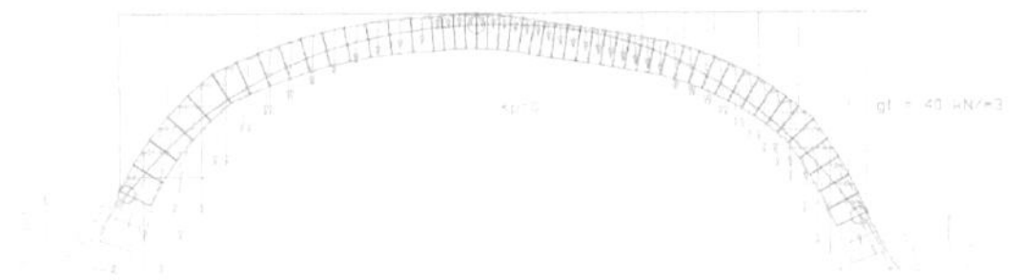

Worst values		Sec.N*
σ e [MPa]	1.08	28
σ i [MPa]	1.08	46

Figure 9.5.86 Diagonal arch 1 – thrust line.

condition, the safety coefficient turned out to be not so high, especially for vault 2, the central and most damaged one. For instance, in this vault, the minimum percentages of compressed section in the two arches in more critical static conditions, i.e. arch 2–3 (Figure 9.5.85) and diagonal arch 1 (Figure 9.5.86), are 36,7 and 29,6, respectively, with safety coefficients 1,468 and 1,184, respectively.

arch 2–3 min % compressed section = 36.7% > 25% safety coefficient = 1,468

diagonal arch 1 min % compressed section = 29.6% > 25% safety coefficient = 1,184

The analyses also show that the safety coefficient with the existing load conditions (*ante operam*) is slightly higher than the one with the design load conditions (*post operam*), as was expected. Nevertheless, the computed strength values of masonry exceed the admissible ones, which normally does not occur with the design load conditions. In fact, the highest compression value in the arches of vault n.2 with *ante operam* load conditions is 5,50 daN/cm² and it exceeds by 59,7% the design strength value in static conditions ($f_{c,d}$ = 3,38 daN/m²). Hence the need, already foreseen in the design phase, to carry out a widespread consolidation of the masonry of the vaults by means of mortar injections with the aim of filling gaps and internal voids, as far as possible, and increasing the masonry strength up to values significantly higher than those emerging from numerical modeling. In conclusion, these preliminary and simplified analyses confirmed the possibility of intervening on the existing vaults without changing their original functioning and geometry, by reducing however the seismic masses in order to get a better dynamic response in the event of an earthquake.

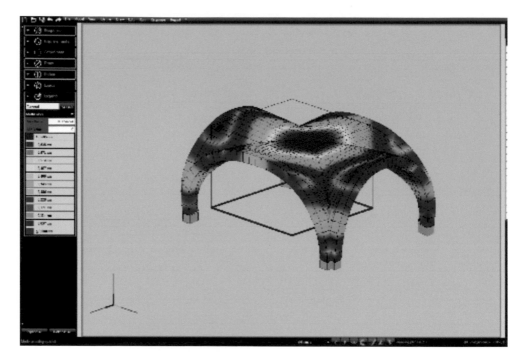

Figure 9.5.87 Deformed vault under vertical loads.

To complete the study on the static and dynamic response of the vaults a non-linear analysis was carried out on vault n.2, the most damaged one. The modeling was carried out with the help of the HiStrA structural code [11], a specialized software for non-linear analyses of arch systems and masonry vaults. The software is able to carry out 3D modeling of different types of arches and domes, even with irregular shapes, and static and dynamic non-linear analyses based on the push-over method. The push-over modeling confirmed that in the presence of only vertical loads (Figure 9.5.87) this vault has an acceptable safety margin which however disappears under dynamic conditions (Figures 9.5.88, 9.5.89) in the presence of which the checks are no longer satisfied. In fact, in this case plastic hinges are formed which lead to the collapse of the vault for values of the seismic action equal to about 30% of the design one.

The modeling thus confirms the design choice of using glass fiber nets embedded in the plaster to give tensile strength to the masonry, so preventing or at least delaying the formation of plastic hinges at the vault intrados and extrados. The use of reinforcement systems made of glass fibers, carbon and basalt nets inserted in inorganic cementitious matrices is one of the latest and most innovative reinforcement and consolidation systems for walls, arches and vaults. The main advantages consist of high mechanical performances, low architectural impact, high durability, easy application and reversibility. Moreover, the introduction of fiber-reinforced mortars with lime-*pozzolana* base guarantees great compatibility with historic mortars. The use of these composite materials in an inorganic

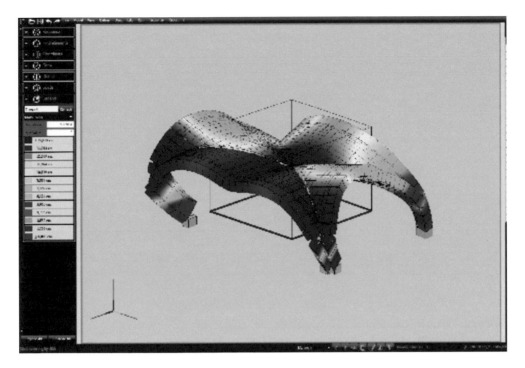

Figure 9.5.88 Deformed vault under x seismic action.

matrix makes it possible to compensate for the deficiencies in tensile and shear strength of masonry and to confer a sort of ductility to the overall behavior of the structure subject to intervention.

The overall behavior of the walls-vaults complex was checked by using the 3D Finite Element code PRO_SAP [12]. For a better and more accurate analysis, 3D nonlinear models of the entire basilica (Figure 9.5.90) and of the narthex in particular (Figure 9.5.91) were created by using more details, more refined meshes and nonlinear finite elements especially in the narthex area. More specifically, the lack of connection of the two internal walls of the narthex with the façades of both narthex and church was modeled by introducing unilateral contact conditions, i.e. internal links acting only in compression. The same was done with the infill walls of the main door and the two side doors of the narthex façade and of the two smaller doors that lead into the aisles. Also the "safety system" for hanging the vaults to the new steel roof structure was modeled and checked with the Pro Sap software.

Here again, checks were carried out with reference to the conditions *ante* and *post operam*. The results concerning the vaults confirm what had already emerged from the linear and non-linear modeling described above; as regards the masonry walls, and in particular the façade of the narthex, the extrados–tie rod system and the steel beams in the place of the filling, together with the improvement of the quality of the masonry of the vaults, allow obtaining a significant reduction of the risk of collapse both with ordinary

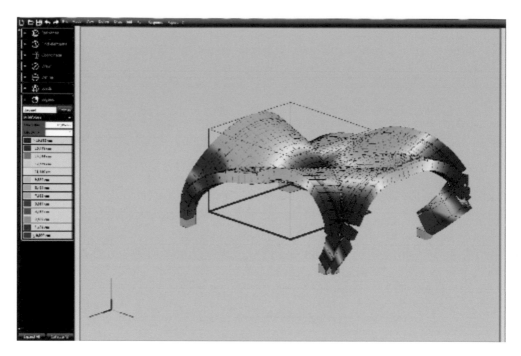

Figure 9.5.89 Deformed vault under y seismic action.

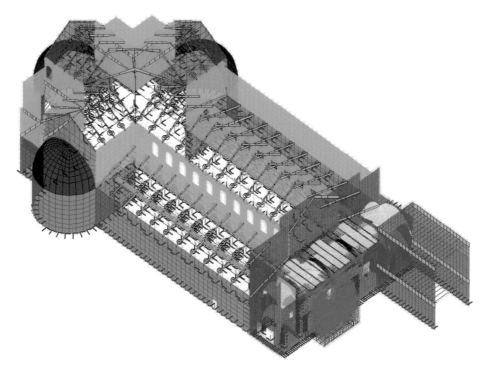

Figure 9.5.90 3D F.E.M. model of the church.

Figure 9.5.91 3D model of the narthex.

vertical loads and under seismic actions. In the consolidated situation, the only areas that remain unverified, and therefore at risk of collapse (but with a considerable reduction in the percentage exceeding the safety limit), are those corresponding to the buffered side.

9.5.2 Anchorage and on-site tests

The diagonal tie rods, the steel beams and the curb are strongly anchored to the masonry structure by means of injections of suitable adhesive mixtures. In particular, for the diagonal tie rods, deep anchorages were made with the Bossong System based on the use of special containment bags and lime mortar. At the end of the works, before tensioning the tie rods up to the design load and laying the new floor, some pull out tests were carried out with a EUROPRESS portable extractor, mod. CMF30N100, to check the efficiency of the injections and the achievement of the planned project resistances (Figure 9.5.92).

Since the anchorage involves two distinct walls (the façade of the church and the one of the narthex) that might have different mechanical characteristics, two tests were carried out, one for each façade. These non-destructive tests, aimed at verifying the ability of the anchors to support the project load, were carried out in four steps:

1 The anchorage was initially loaded with a percentage of the design load (about 46%), and the constancy of the test pressure, which denotes the absence of anchorage failures, was monitored for about 15 minutes.
2 Then the pressure was increased up to the design load (2700 daN), and the constancy of the test pressure was monitored for another 15 minutes.

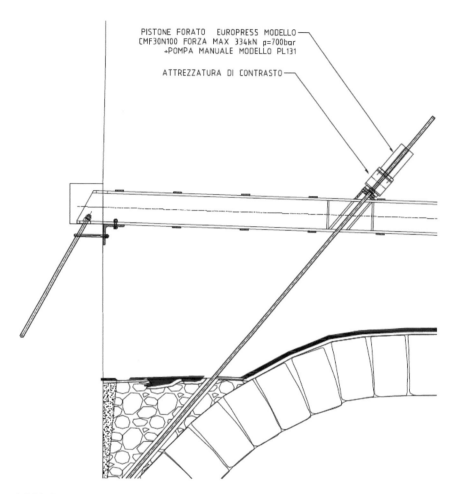

PISTONE FORATO EUROPRESS MODELLO
CMF30N100 FORZA MAX 334kN p=700bar
+POMPA MANUALE MODELLO PL131

ATTREZZATURA DI CONTRASTO

Figure 9.5.92 Diagonal tie rod with EUROPRESS portable extractor in pull out test.

3   The pressure was increased by 20% of the design load, and again the constancy of the test pressure was monitored for another 15 minutes.
4   The pressure was returned to zero.

All tests were successful.

9.6   The Armenian door

M. Bacci and S. Sarmati

9.6.1   Main features

In 1187, when Jerusalem and most of Palestine, including Bethlehem, were recaptured by Saladin's Egyptian army, the Frankish population fled to the Crusader-hold strongholds on

Figure 9.6.93 The upper part of the Armenian door – west front.

the coast and the Latin church lost control of the major holy sites. As an outcome of basically geo-politic considerations, Saladin and his successors refrained from restoring the hegemonic role the Greek church had exerted prior to the Crusades and engaged in granting special rights to other Oriental Christian denominations, including non-Chalcedonian Copts, Ethiopians and Armenians, whereas among the Orthodox, Georgians were encouraged to take control of many important *loca sancta*.

In so doing, the court of Cairo aimed to restrict Byzantine interference in Palestinian affairs, and this attitude was even reinforced when Constantinople fell into Frankish hands in 1204. On the other hand, good relations were established with another important actor in Middle Eastern politics, the Armenian Kingdom of Cilicia, in what is presently south-eastern Turkey. The Armenian door in the narthex of the Nativity Church (Figure 9.6.93) is the most prominent witness to such historical circumstances: contrary to Islamic consuetude, that prevented the *dhimmi* – the Christians living under Muslim rule – from enlarging, decorating or even repairing their religious buildings. The embellishment of the entrance from the narthex into the nave was authorised and even encouraged by Ayyubid authorities just 40 years after their rule on Palestine had been restored.[1]

Two inscriptions, one in Arabic (Figure 9.6.94) and one in Armenian (Figure 9.6.95), were carved on the upper margins in the aim to publicly record the names of the people involved in the making of this sumptuous addition to the church décors.

The choice of a bilingual inscription is itself meaningful, since it works in much the same way as the monumental record, displayed in the altar space in 1169, where the names of the designer – Master Ephraim – and the promoters of the mosaic decoration – King Amalric I of

Figure 9.6.94 Left upper part with inscription in Arabic.

Figure 9.6.95 Right upper part with inscription in Armenian.

Jerusalem, Emperor Manuel of Byzantium and Bishop Ralph of Bethlehem – were mentioned in elegant Greek and Latin verses, the languages of the two most prominent communities in Crusader Palestine and of the international powers supporting them. Latins and Greeks entering the church were now made aware that the political situation had substantially changed: even if they could not read, it was enough to acknowledge the alien graphic shape of the displayed scripts to understand that rulership belonged to others in that holy place.

The Arabic inscription, displayed on the left side of the upper margin, states that the work was finished in the month of Muḥarram 624 H – corresponding to December 22, 1226, through January 20, 1227 – during the rule of Saladin's nephew, the Sultan of Damascus al-Mu'aẓẓam 'Īsā:

> *This door was completed with the help of God, may He be exalted, in the days of our Lord, the Sultan al-Malik al Mu'aẓẓam, in the days of [the month of] Muḥarram in the year 624.*[2]

On the right side, the Armenian inscription indicates that the wooden door was carved by the priests Abraham and Aṙak'el in 1227, that is at the very beginnings of the reign of King Het'um I (Het'own) of Cilicia (1226–1270):

> *In the year 676 of the Armenian era this door of [the church of the] Holy Virgin was made by the hands of priest Abraham and priest Aṙak'el during the reign of King Het'um, son of Konstande [Konstandin]. May God have mercy on the workers.*[3]

It is likely that, on the analogy of the dedicatory inscription of the mosaics, the hint at contemporary rulers was meant to celebrate the latter's merits as promoters of church embellishments. For Het'um, this may have been part of a wider strategy to assert not only the rights of the Armenian church on the holy sites but also his personal piety and leading role as Christian ruler in the wider Near Eastern region. On the other hand, the mention of the Sultan of Damascus was probably far from fortuitous, and it is unlikely, as Moshe Sharon has suggested,[4] that he may have been completely unaware of the monumental inscription in Bethlehem, especially if one considers that the site of Christ's birth was regarded as worship worthy by many Sunni Muslims and that the Sultan himself, whose name was 'Īsā (Jesus in Arabic), may have been interested in having his name exhibited on the main entrance to the Nativity Church. In any case, the double inscription bears witness to both the good political relations between the Ayyubid government of Syria and the Kingdom of Cilicia in the second quarter of the 13th century and the privileged role played by the Armenian community in Bethlehem in that period.[5] Ironically, everything was to change just two years later, when the German Emperor, Frederick II, made an agreement with the Sultan that led to the restitution of the Christian holy sites to the Latin church.[6]

In spite of later restorations, much of the original door is still in a relatively good state of preservation, as the recent conservative works have shown. As we will see below, it is likely that what came to us is only a portion of the original 13th-century structure, which was conceived of as a double-leaved door, as is clearly witnessed, in 1422–1423, by the pilgrim John Poloner, who hints at *bifores*, "shutters", which, he noted, were "made out of cypress wood and carved with multiple ornaments".[7] This is also clearly indicated by the presence of a batten, whose upper part is still extant and partly preserves its original carvings, consisting of vegetal interlaces, horizontal listels and a cross pattée.[8] An elegantly carved decoration was used also for the two shutters, which were structured according to an analogous pattern, with four rectangular panels delimited, on each side, by a cross-shaped, protruding element, which, as the recent restorations revealed, were also originally embellished with inlaid work.

The upper panels (Figures 9.6.94, 9.6.95) display crosses of two different types: on the left, they are in the form of crosses potent, whereas those on the right have flaring arms terminating in projecting, trilobed knobs at each tapering end.

Furthermore, the hill of Golgotha on which they stand is evoked, in the former case, by a stepped base and, in the latter, by a curvilinear shape. In all panels, the body of the cross is decorated with ornamental motifs: to the left, the margins are embellished with a continuous band of dotted ornaments, whereas the internal parts are filled by scrolls with sprouting two-leaved flowers that intertwine with square grid-shaped ornaments (Figure 9.1.96). The latter are absent on the right crosses, which display a thoroughly vegetal décor with leaves and buds of different shapes (Figure 9.6.97).

Analogous is the treatment of the background, where large bunches of grapes and some smaller leaves sprout out of the intertwining scrolls in a rigorously symmetrical disposition. On the other hand, the lower panels (Figure 9.6.93) – one of which is lost – are embellished with aniconic motifs, consisting of interlacing scrolls giving shape to roundels with sprouting flowers composed of two hastate leaves and a vertically elongated bud. The protruding, cross-shaped frame is carved with an arabesque décor consisting of six interlacing spiralling stems with elongated, crenate leaves ending in a curling edge (Figure 9.6.93). The carvings on the upper vertical elements stand out for a significant detail: at the meeting point of the innermost spirals are displayed two flabellate or fan-shaped flowers. On the lower, horizontal elements of the frame the same position and function are given to half-moon motifs.

Figure 9.6.96 Cross on the left panels.

Figure 9.6.97 Cross on the right panels.

It can be wondered what we can really infer, on historical grounds, from the analysis of the multifarious ornamental motifs displayed in this artwork. Undoubtedly, the practice of contributing to the enhancement of a church decorum by donating a sumptuously carved wooden door was particularly common in Medieval Armenian culture, as is witnessed by a relatively significant bulk of preserved materials, all belonging to the Sub-Caucasian area.[9] Furthermore, the Armenian inscription clearly indicates that Father Abraham and Father Aṙak'el were materially responsible for the carving of the wooden surface. Nothing is known of these two priests, but it can be assumed that the absence of any hint at their foreign provenance may indicate that they rather belonged to the Armenian community of Palestine than to Cilicia or Greater Armenia.

The forms displayed on the door indicate, in any case, that preference was given to patterns of ornamental art widespread in the production made in contemporary Ayyubid lands for both Islamic and Christian communities. The Arabic inscription makes use of the elegant *nakshī*-type cursive calligraphy which had first been introduced in Syria during the rule of Nūr al-Dīn (1146–1174).[10] The display of complex interlacing spirals with elongated, hastate and crenate leaves, stylised flowers, dotted borders, and half moons is reminiscent of contemporary Islamic wood carving, such as, for example, the two sculpted panels from 13th century Egypt in the Benaki Museum in Athens.[11]

On the other hand, some details point to an intentional display of features specifically associated with Armenian arts, though interpreted in a very idiosyncratic way. The visual emphasis laid on elaborate panels reserved for the image of the holy cross is deeply rooted in

Figure 9.6.98 Stone slabs known as *khač'kars*.

Armenian tradition and seems to be reminiscent of the image-type associated with the stone slabs known as *khač'kars* (Figure 9.6.98).

The presence of grapes in the four panels is instrumental to convey Eucharistic symbolism: rooted in early Christian tradition, this detail had been frequently regarded by Armenian artists as commonplace in the decoration of church exteriors[12] and had become widespread in the illuminated frontispieces of Cilician Gospel books since the 12th century.[13] From the same time it also occurs in the carved decoration of *khač'kars*, as is indicated, e.g. by a late 12th or early 13th century formerly in Havuts' Tar Monastery and now in the History Museum of Armenia in Yerevan.[14] Most interestingly, the same motifs are given visual prominence in the earliest, Armenian Palestinian *khač'kars*, from the mid-12th century, preserved in the Monastery of Saint James in Jerusalem. In general terms, the latter include both crosses bearing strong iconographic connections with contemporary works from Great Armenia (as concerns, for example, the leaves sprouting out of the lower end, hinting at the symbolism of the Tree of Life), and others which, as in the Bethlehem door, seem to intentionally lay emphasis on the locative character of the Crucifixion by substituting the usual cross-supports (such as small columns, diminutive stepped bases or circles) with a stylised, elevated, stepped or curvilinear shape evoking the hill-like outline of Mount Golgotha.[15] In this respect, this iconographic solution is paralleled also by some extant carved stone crosses made for Syriac-rite communities in the Ayyubid Near East[16] and a number of 13th-century Coptic manuscripts, which also share a preference for he cross potent filled with vegetal and geometric interlaces on a high, stepped base.[17]

In its present state, the door is the outcome of a restoration that took place sometimes in the late 16th or early 17th century. In the sixth century, the church façade had been provided by Justinian's architects with three monumental entrances, which, as witnessed by the Frenchman Pierre Mesenge,[18] were still extant in 1507: both the northern door and the one to the south, being accessible from the only remnant of the ancient atrium – the so-called School of Saint Jerome incorporated in 1723 into the Armenian convent[19] – were then described as solid and magnificent wooden doors, even if they were never opened out of fear that Bedouins and local Muslims might freely enter and cause some violence. The only access to be used was by then the central entrance, whose size had been reduced for security reasons and probably corresponded to the still-visible outline of a pointed arch.[20] The latter's further reduction to the small rectangular opening known as the Door of Humility probably took place after the Ottoman conquest of Palestine in 1516. The situation worsened, since soldiers and officers, including the *qadi*, were used to station in the nave during the day. Given that the latter was *de facto* appropriated for other purposes, the access to it lost its importance, and Western pilgrims visiting Bethlehem became therefore accustomed to access the holy grotto through the Franciscan convent.[21] It can be assumed that, in order to prevent these people from entering the church with horses or camels, the side entrances were walled up, whereas the central door was reduced to its present size and provided with a small but thick wooden door reinforced with iron.[22]

Since the Flemish traveller Jan Zvallart, in 1586, compares the diminutive measures of the Door of Humility with the imposing proportions of the narthex one, said to be "big and high, carved in the ancient way",[23] it can be assumed that, by then, the latter's reshaping had not yet taken place. Ten years later, as reported by the Flemish Jan Kootwijck (Cotovicus), the outward appearance of the door had been altered:

> *Upon entering one finds first a very large porch with vaults. And once you have gone through it, you will see wooden shutters, being ca. twelve feet high and eight large. The latter are never or seldom opened, since they are blocked with bars intersecting like a cross, and there is only a narrow, reduced open wicket, through which one can access the church. But they are very old shutters, made of cedar in the Syrian way and with inscribed Arabic letters.*[24]

Further information is given, in 1599, by the Italian pilgrim Aquilante Rocchetta, who writes:

> *Letter B [in Aquilante's plan of the church] corresponds to the second entrance, which includes a big and high door: it is sumptuous enough and carved in the old fashion, out of cedar wood, as it is used in Syria, with many Arabic letters on it. This one is also never opened: there is only a small opening of three spans in width and five in height, to prevent horses from entering. The latter's door is made of a half-span thick wood, with two wooden bars superimposed in the way of Saint Andrew's cross, so that it may be protected from the hands of enemies. This smaller door is covered with iron plates to make it fire-proof.*[25]

On the one hand, these two texts clearly indicate that important restorations took place in the last years of the 16th century, and, on the other, they point out that the alterations did not go so far as to prevent the full opening of the two shutters. The date 1626 incised on the rear and some other clues suggest, anyway, that the door underwent further alterations still in the following decades. Whereas most of the panels are richly carved with complex interweaving motifs that originally looked much more like intarsia than as a wooden imitation of lacework, as one would infer from their present-day appearance, it is striking to observe that the vertical elements of their frame were simply incised with rather linear vegetal scrolls.

Zehava Jacoby has proposed to interpret the latter as a sort of preparatory sketch or *sinopia*, which was made visible at some point, when the superimposed carved panels were lost.[26] Nevertheless, this is contradicted by the fact that the vertical beams are flush with the nearby horizontal ones and that the latter were carved out of the wooden surface and not fixed to it. Furthermore, the scrolls with large crenate leaves or palmettes do not belong to the ornamental repertory used in the rest of the panels and bear closer affinities with motifs used in Armenian book illumination from the 14th century onwards. Such motifs were widely circulating in diaspora communities not only via manuscripts but also in the form of model books and collections of drawings, such as the one preserved in the Library of the Mekhitarist monastery of San Lazzaro in the Venetian lagoon, which is thought to have originally been produced in the Lake Van area in the late 15th century and integrated with new images in 16th-century Constantinople.[27]

It is therefore much more probable that the incised beams were added in a later phase, probably in the aim to repair or consolidate the general structure. The presence of an Armenian inscription on the vertical beam to the right indicates that the community was responsible for this restoration. It reads:

Lord God, have mercy of Alic, for she is like a rose.[28]

Alic is the Armenian equivalent to the Western name *Alix*, which became popular in Cilicia from the 12th century onwards. Since it became an attribute of many female members of the Rupenid and Hethumid dynasties, it was frequently used also in non-courtly circles in later periods.[29] The text should be interpreted as the invocation to God by a lady who aimed to manifest her piety on account, most probably, of her involvement in the financial support of the restoration works, since it would be hardly conceivable that a private woman may have been granted the right to exhibit her name in such a prominent position, had she not been perceived as a special benefactor.

Some considerations can help us to determine the approximate date of this intervention. It is most unlikely that the Armenians may have been authorised to make repairs to the door during the Mamluk period (1250–1517), and especially from the mid-14th century onwards, when the Franciscans were granted the *praedominium* in the basilica and were therefore responsible for all maintenance works, such as those of the roof made in the 1480s. There are scarce details about their presence in Bethlehem before the early 17th century. Armenians are known to have celebrated in the north transept since at least the 14th century,[30] and in 1494 Pietro Casola wrote that this community including men, women and children lived there "without shelter", so possibly somewhere in the ruined buildings joining the basilica.[31] In 1512, Jean Thenaud mentions Armenian monks living in the southern church annexes,[32] whereas Kootwijck, at the end of the 16th century, makes clear that the cells were also inhabited by Copts and Syrians,[33] and the plan of Bernardino Amico (1609) indicates that the latter were mostly on the upper floor, over the structures of the so-called School of Saint Jerome, which was by then used as a stable. Nevertheless, most of the structures adjoining the southern side of the church were by then disused or appropriated for profane activities, such as the weaving of cloths.[34]

The situation changed in 1621, when, in keeping with his strategy to revitalise Armenian pilgrimage to the Holy Land,[35] Patriarch Krikor Parontēr purchased the whole complex of structures to the south of the basilica and renovated them with the construction of cells, a pilgrims' hospice, a refectory and a church dedicated to the Virgin Mary.[36] From that period onward, the narthex came to be used also as the main entrance to the new Armenian monastery. It can be assumed that the final restoration of the door took place, with the financial support of a lady named Alic, on the occasion of these renovations: its shape was altered in such a way as to thoroughly prevent its full opening, and some deteriorated beams were substituted with new ones, bearing simple, incised ornaments.

9.6.2 Interventions of consolidation

The 13th-century wooden door in the narthex is 4.56 metres high and 3.4 metres wide. It was assumedly altered, repaired and changed many times throughout its lifetime. The original door was made by walnut and cypress wood, with a double layer of timber in the outer and inner boards. The outer side-boards are inlaid and carved with a double inscription. The inner side-boards are made of smooth panels. In the upper part of the door the original panels on the outside are carved, whereas the original panels on the inside are not decorated. Both original outer and inner panels are fitted into wooden frames, formed by four interlocked pillars. Originally, the door was a double-leaved one and each part was 1.57 metres wide and opened inward. The two big wooden hinges, which allowed the door opening, are still present, located in holes in the stone lintel. The door owes its current appearance to the restoration made in the 17th century: the lower part, which was probably seriously damaged, was reshaped, and the access to the church was reduced in size. The original structure was completely modified by building a lower panel with a little door to allow entry into the basilica.

The recent restoration brought back to light the inner panel of the door, which had been completely covered with wooden beams. Furthermore, some wooden elements were removed to check their state of conservation. In the course of this removal process it was possible to establish that the horizontal beams located on the inner part of the door were not original and were probably due to a previous restoration, made to reinforce the structure and block the door. These horizontal beams were thoroughly superimposed onto the inner panels of the door. Deemed to be lost, they were recovered thanks to careful restoration work. The date 1626 discovered on a beam on the rear side probably hints at the year of the restoration. By removing the beams, it was also possible to recover the original structure of the door on the inner part, bringing back to light the original pillars and their cross-shaped structure, which was entirely covered. On the outer side of the door, two original carved panels were discovered by removing some wooden boards; placed on the right and the left side, they had remained hidden for centuries. Thanks to these removals it was possible to recover the original appearance of the upper part of the door (Figure 9.6.99).

The state of conservation of the wooden structure was fairly good, despite tampering and reconstructions. The door was deformed because of a disconnected anchoring system and had lost its functionality due to the attacks of xylophage insects. Some elements, probably made of sapwood, were seriously damaged and had very deep cavities, caused by original

Figure 9.6.99 The door after the restoration.

material pulverisation and loss (Figure 9.6.100). The surface was covered with dust and candle smoke deposits. Moreover, there was a strong chromatic distortion due to the presence of a thick compact brownish grey coating, caused by the oxidation of old waxes and animal glues that had been applied during previous restorations and whose ageing and resulting polymerisation process had caused a change in the chemical structure.

Prior to the restoration the team promoted a detailed photographic campaign of the entire artefact with normal light, grazing light and the UV fluorescence technique, in order to discover remakes on the carved panels or any colour marks. This procedure confirmed the absence of any kind of colour or pigment on the wooden surface. The dust on the surface was removed with a micro-cleaner and with small brushes. The wood disinfestation was made with a liquid biocide, Permetar©, whose active ingredient is Permethrin. The product was applied with a brush to let the insecticide go deeper into the most affected parts

Figure 9.6.100 The wood seriously damaged.

and impregnate the material through a spreading process. The disinfestation was generally made by using both physical factors (heat, gamma radiation, microwaves) and chemical means (toxic gas fumigation, treatments with insecticides). The use of just physical means would not be enough to prevent possible insect attack. So it was decided to use a solution of insecticides which remains in the wood so as to prevent new infestations, as well as insect spawning. The wooden structure had been seriously damaged by woodworms. Tests were carried out in order to choose a stabilising agent with a strong penetrability, such as to provide the wooden structure with a good mechanical strength. It was decided to apply a 5% solution of an acrylic product, the ParaloidB72, in acetone. The stabilising agent was applied using brushes and inserting syringes in the worms' holes, in repeated cycles of alternate wetting and drying. The missing pieces of the door were replaced with the same kind of ancient wood by applying seasoned wood pieces with wooden pins and following the pattern of the fibres. After the consolidation and the biocide treatment, tests (Feller's solubility test) were carried out in order to choose the best solvent or the best solvent mixture to remove wax deposits of oxidised varnishes from the wooden surface. After several solubility tests, dirt was removed by using brushes and cotton swabs, with an absolute alcohol solution in different degrees of concentration (absolute ethyl alcohol and distilled water; Figure 9.6.101).

The cleaning process allowed us to identify the original colour of the wood, bringing back to light its natural grave and the two colours of the carved surfaces. The restoration work has given a new dignity to the door, thanks to the elimination of the improper elements and the recovery of important parts of the decoration. However, it was not possible to restore the original functionality of the door because of the 16th-century restoration that had completely modified the door's structure.

Figure 9.6.101 The cleaning test.

9.7   Maintenance

C. Alessandri and M. Martinelli

For general considerations on the maintenance plan of structures, the reader is referred to Section 10 of Chapter 3.

The consolidation intervention concerned the three cross vaults of the central part of the narthex.

This Maintenance Plan [7] concerns the new steel structures added to the existing vaults and the reinforced plasters that cover the surfaces of the vaults.

9.7.1   Steel structure

User manual

It consists of all the steel elements (beams, corrugated sheets, threaded bars, bolts, cables) which form the consolidation structure of the existing vaults. The aim of this structure is to transfer to the walls of the narthex the loads coming from the floor above the vaults (self-weight of the floor and live loads), the thrusts exerted by the vaults (bearing their own self-weight only) and those produced by seismic actions. Moreover, the vertical thin tie rods connecting the vaults to the horizontal beams contribute to prevent the collapse of the vaults in the case of exceptional and/or unexpected events and/or in the presence of vertical seismic actions. The durability of all these elements is guaranteed both by the use of stainless steel and by a galvanizing process extended to all the elements that are not made of stainless steel.

All these elements are properly used if the operating loads are kept constantly within the design values.

Maintenance manual

> *Minimum level of performance*: resistance to design stresses.
>
> *Detectable anomalies*: bubbles or cracks of the protective layer with risk of corrosion (chemical and biological causes). Water infiltration, excessive deformation of the elements, loosening of the bolts, relaxation of the tie rods, cracks in the supporting masonry walls.
>
> *Checks*: they can be carried out through a visual inspection aimed at identifying the damaged element.
>
> *Frequency of the checks*: every year.
>
> *Executions of the checks*: they can be performed by the user.
>
> *Interventions*: application of anti-rust products and restoration of the protective layer. If some of the structural anomalies listed above occur, a specialized professional must be called in to evaluate their actual severity and danger. If necessary, safety measures and static checks must be provided for, followed by possible consolidation works.
>
> *Frequency of the interventions*: whenever it is deemed necessary.
>
> *Execution of the interventions*: they can be performed by the user, if they concern the restoration of protective layers. Safety and consolidation interventions must be carried out by specialized personnel on the basis of a project drawn up by a qualified professional.

9.7.2   Reinforced plaster

User manual

It provides an adequate degree of finishing to the surfaces of the vaults and increases the strength and ductility of the masonry structure. The plaster reinforcement consists of a high-resistance net of basalt fibers, highly compatible with the existing masonry. Such a compatibility contributes to guarantee the durability of the plaster, along with the high quality of the materials used.

The plaster works properly if the operating loads are kept constantly within the design values.

Maintenance manual

> *Minimum level of performance*: resistance to design stresses and maintenance of the required degree of finishing.
>
> *Detectable anomalies*: cracks and micro-cracks, plaster detachments, moisture stains and water infiltration.
>
> *Checks*: they can be performed through visual inspections, to be performed both on the extrados and on the intrados of the vaults. They should help identify the occurrence of damage.
>
> *Frequency of the checks*: every year.
>
> *Execution of the checks*: they can be performed by the user.

Interventions: if cracks or micro-cracks occur, stuccoing and restoration must be carried out with the same (or compatible) material as the existing one. Sources of water infiltration must be previously identified and eliminated, then the restoration will be carried out with the same (or compatible) products. In the case of serious cracks or detachments, a specialized professional must be called in to assess their actual severity and danger. If necessary, safety measures and static checks must be provided for, followed by possible consolidation works.

Frequency of the interventions: whenever it is deemed necessary.

Execution of the interventions: these must be performed by specialized personnel. Safety and consolidation interventions must be carried out by specialized personnel on the basis of a project drawn up by a qualified professional.

Notes

1 See, in general, M. Bacci, *The Mystic Cave: A History of the Nativity Church in Bethlehem*, Rome, Brno 2017, pp. 207–210.
2 M. Sharon, *Corpus Inscriptionum Arabicarum Palaestinae*, Volumes 1–5 in continuation, Leiden, 1999–2013, 2, pp. 184–187.
3 H. Vincent & F.-M. Abel, *Bethléem. Le sanctuaire de la Nativité*, Paris, 1914, pp. 184–185; R.W. Hamilton, *The Church of the Nativity, Bethlehem: A Guide*, Jerusalem, 1947, pp. 48–49; K. Hintlian, *History of the Armenians in the Holy Land*, Jerusalem, 1989, pp. 42–43; G. Grigoryan, *Royal Images of the Armenian Kingdom of Cilicia (1198–1375) in the Context of Mediterranean Intercultural Exchange*, unpublished Ph.D. dissertation, University of Fribourg, 2017, p. 74, English translation after this latter work, Fribourg.
4 Sharon, *Corpus Inscriptionum Arabicarum Palaestinae*, pp. 186–187.
5 J.G. Ghazarian, *The Armenian Kingdom of Cilicia during the Crusades: The Integration of Cilician Armenians with the Latins, 1080–1393*, Richmond, 2000, p. 60; C. Mutafian, Prélats et souverains arméniens à Jérusalem à l'époque des Croisades: légendes et certitudes (XIIᵉ-XIVᵉ siècles). *Studia Orientalia Christiana Collectanea*, 37 (2004), 109–151, here 139; C. Mutafian, *L'Arménie du Levant (XIᵉ-XIVᵉ siècle)*, Paris, 2012, p. 713.
6 Bacci, *The Mystic Cave*, pp. 209–211.
7 J. Poloner, *Descriptio Terrae Sanctae* (1422), T. Tobler (ed.) *Descriptiones Terrae Sanctae ex saeculo VIII. IX. XII. et XV.*, Leipzig, 1874, pp. 225–281, here p. 248.
8 The most thorough study of the Armenian door is provided by Z. Jacoby, The medieval doors of the Church of the Nativity at Bethlehem. In: Salomi, S. (ed.) *Le porte di bronzo dall'Antichità al secolo XIII*, Rome, 1990, pp. 121–134. Cf. also Vincent and Abel, *Bethléem*, pp. 184–185; Hamilton, *The Church of the Nativity, Bethlehem*, pp. 48–49.
9 V. Nersessian, *Treasures from the Ark: 1700 Years of Armenian Christian Art*, Los Angeles, 2001, pp. 139–143.
10 Y. Tabbaa, *The Transformation of Islamic Art during the Sunni Revival*, Washington, DC, 2001, pp. 60–63.
11 M. Moraitou, 37–38 Deux panneaux sculptés. In: Delpont, E. (ed.) *L'Orient du Saladin. L'art des Ayyoubides*, exhibition catalogue (Paris, Institut du monde arabe), Paris, 2001, p. 47.
12 A. Kyurkchyan & H.H. Khatcherian, *Armenian Ornamental Art*, Yerevan, 2010, pp. 91, 93–95, 97.
13 For notable examples, cf. Yerevan, Matenadaran, Ms. 7347, AD 1166, fols 13 and 104 (S. Der Nersessian, *Miniature Painting in the Kingdom of Cilicia from the Twelfth to the Fourteenth Century*, Washington, DC, 1993, pp. 4–6, Figs 9–10), and Tokat, AD 1173 (Der Nersessian, *Miniature Painting in the Kingdom of Cilicia from the Twelfth to the Fourteenth Century*, Figure 13).
14 C. Maranci, Cat. 38. Khachkar. In: Evans, H.C. (ed.) *Armenia: Art, Religion, and Trade in the Middle Ages*, exhibition catalogue, New York, 2018, p. 92.
15 H. Katchadourian & M. Basmadjan, *L'art des khatchkars, les pierres à croix arméniennes d'Ispahan et de Jérusalem*, Paris, 2014, pp. 41, 111–123, 384–385.

16 B. Snelders, *Identity and Christian-Muslim Interaction: Medieval Art of the Syrian Orthodox from the Mosul Area*, Leuven, 2010, pp. 325–328 and pls 34, 48, 58, 64, 65.

17 J. Leroy, *Les manuscrits coptes et coptes-arabes illustrés*, Paris, 1974, pp. 57–61.

18 P. Mesenge, *Livre et exortation* (1507), M. Muller (ed.) *Livre et exortation pour esmouvoir les chrestiens d'aller visiter le Sainct Sepulchre de Nostre Seigneur en Hierusalem et les aultres lieux sainctz en la Terre Saincte, par Pierre Mesenge. Transcription et édition annotée, Amiens, Bibl. Mun., Fonds Lescalopier 099 C (5216)*, Master thesis, University of Fribourg, 2016, p. 95, Fribourg: "Aprés ladicte gracieuse reception, entrasmes en ladicte eglise par ung bien petit huys, combien qu'il y ait deux plus grandes portes de boys bien ferrees et bien magnificques. Mays *ad* ce que j'ay peu entendre par les freres qui nous conduysoint, on ne les ouvre point pour crainte que les gens du paÿs n'y entrent en grand nombre et n'y facent quelque violence".

19 Hintlian, *History of the Armenians in the Holy Land*, p. 43.

20 The latter has been tentatively interpreted as dating from the Crusader era by D. Pringle, *The Churches of the Crusader Kingdom of Jerusalem: A Corpus*, Volume 1, Cambridge, 1993–2009, p. 140, but it is difficult to imagine that the Latins may have felt the need to reduce the dimensions of the central entrance without altering the side ones. Hamilton, *The Church of the Nativity, Bethlehem*, p. 47, suggested that the pointed arch dated from the thirteenth century or later.

21 M. Bacci, Old restorations and new discoveries in the Nativity Church, Bethlehem. *Convivium*, 2(2) (2015), 36–59, here 46; Bacci, *The Mystic Cave*, pp. 240–241.

22 First hints at the diminutive dimensions of the outer door occur in B. Morosini, *Viaggio ai luoghi santi* (1514), da Civezza, M. (ed.) *Le missioni francescane in Palestina*, Florence, 1891, p. 236; O. Prefát, *Cesta z Prahy do Benátek a odtud potom po moři až do Palestyny*, K. Hrdina (ed.) *Cesta z Prahy do Benátek a odtud potom po moři až do Palestyny, to jest do krajiny někdy Židovské, země Svaté, do města Jeruzaléma k Božímu hrobu, kterakžto cestu s pomocí Pána Boha všemohúcího šťastně vykonal Voldřich Prefát Vlkanova leta Páně MDXXXXVI*, Prague, 1947, p. 168. Cf. the detailed description of the wooden door in J. Goujon, *Histoire et voyage de la Terre-Sainte*, Lyon, 1671, pp. 268–269.

23 J. Zvallart, *Il devotissimo viaggio di Gierusalemme fatto e descritto in sei libri dal signor Giovanni Zuallardo, cavaliere del Santissimo Sepolcro di N. S. l'anno MDLXXXVI*, Rome, 1595, p. 207. Cf. also L. Vulcano, *Vera, et nuova descrittione di tutta Terra Santa, et peregrinaggio del Sacro Monte Sinai, compilata da verissimi autori*, Napoli, 1593, p. 79r.

24 J. Kootwijck [Iohannes Cotovicus], *Itinerarium hierosolymitanum et syriacum*, Antwerp, 1619, p. 227: "Ingredientibus primo occurrit porticus amplissima, arcuato opere: quam transgressis valvae sese offerunt ligneae, altitudinis circiter pedum duodecim, latitudinis octo, quae nunquam aut raro aperiuntur, clathris in Crucis modum sese intersecantibus, obiectis, unico angusto ac depresso patente ostiolo, per quod ad delubrum transitur. Sunt autem valvae vetustissimae, ex cedrinis lignis, Syriaco opere, literis Arabicis insculptis".

25 A. Rocchetta, *Peregrinatione di Terra Santa, e d'altre provincie*, Palermo, 1630, p. 257: "La littera B è la seconda entrata, nella quale è una porta grande, et alta, assai magnifica, e lavorata all'antica, di legno di cedro, come s'usano nella Soria, con molte lettere arabiche scolpite. Questa similmente non s'apre mai, ma vi è solamente un portello largo tre palmi, et alto cinque, acciò non vi possano entrare cavalli, la porta è di legno grosso mezzo palmo, con due sbarre di legno l'una sopra l'altra acciò si possa difendere dalle mani de' nemici, il qual portello è coverto di lande di ferro per potere resistere al fuoco".

26 Jacoby, The medieval doors of the Church of the Nativity at Bethlehem, p. 128.

27 S. Der Nersessian, *Études byzantines et arméniennes*, Louvain, 1973, pp. 665–672 (here 666), 673–681, and figs 418–419; R.W. Scheller, *Exemplum: Model Book Drawings and the Practice of Artistic Transmission in the Middle Ages (ca. 900–ca. 1470)*, Amsterdam, 1995, pp. 400–412.

28 I am obliged to Dr Gohar Grigoryan (Fribourg) for deciphering and translating this inscription, which was made visible by the recent restorations. Meanwhile, the inscription has been also mentioned by P. Donabédian, *L'autre porte de la Nativité. Terre Sainte*, 659 (January–February 2019), 12–14, here 14. The same author proposes that the door may have been originally located in another place, possibly a chapel in the nearby Armenian monastery. Nevertheless, this hypothesis is not corroborated by extant textual and material evidence.

29 W.H.R. de Collenberg, *The Rupenides, Hethumides and Lusignans: The Structure of the Armeno-Cilician Dynasties*, Lisbon, 1963, pp. 13, 19, 23, 26, 29, 30, 37 (Rupenid dynasty), 13, 15, 16, 20, 32, 36, 37 (Hethumid dynasty).

30 Bacci, *The Mystic Cave*, pp. 214–215.
31 P. Casola, *Viaggio* (1494), A. Paoletti (ed.) *Viaggio a Gerusalemme di Pietro Casola*, Alexandria, 2001, p. 207.
32 J. Thenaud, *Voyage d'outre-mer* (1512), C. Schefer (ed.), Paris, 1884, p. 93.
33 Kootwijck, *Itinerarium hierosolymitanum et syriacum*, 1619, p. 238. Cf. also G. Sandys, *Sandys Travels*, London, 1673, p. 139: "Arminians, who inhabit within on the right hand of the entrance".
34 B. Amico, *Trattato delle piante et imagini de i sacri edificii di Terrasanta*, Rome, 1609, map 1, numbers 3 (disused spaces), 8 (flight-of-steps to the Armenian cells), 27 (space for the weaving of cloths).
35 R. Ervine, Changes in Armenian Pilgrim attitudes between 1600 and 1857: The witness of three documents. In: Stone, M.E., Ervine, R.R. & Stone, N. (eds.) *The Armenians in Jerusalem and the Holy Land*, Leuven, 2002, pp. 81–95, here pp. 82–83.
36 M. Aławnowni, *Haykakan hin vank'er* [Old Armenian Monasteries], Jerusalem, 1931, pp. 246–247; T.H.T. Sawalaneants, *Patmut'iwn Erusaghēmi*, Jerusalem, 1931, p. 574; A.K. Sanjian, *The Armenian Communities in Syria under Ottoman Domination*, Cambridge, MA, 1965, pp. 142–143; Khatchadourian and Basmadjan, *L'art des khatchkars, les pierres à croix arméniennes d'Ispahan et de Jérusalem*, p. 155. For a contemporary Western record dating from 1630, cf. E. Roger, *La Terre Saincte, ou Description topographique très particulière des Saincts Lieux et de la Terre de Promission*, Paris, 1664, p. 204.

References

[1]   Bacci, M., Bianchi, G., Campana, S. & Fichera, G. (2012) Historical and archaeological analysis of the Church of the Nativity. *Journal of Cultural Heritage*, 13, Elsevier, e5–e26.

[2]   Fichera, G., Bianchi, G. & Campana, S. (2016) Archeologia dell'Architettura nella Basilica della Natività a Betlemme, pp. 1567–1589, in Olof Brandt – Vincenzo Fiocchi Nicolai (a cura di) *Acta XVI Congressus Internationalis Archaelogiae Christianae Romae* (22–28.9.2013). *Costantino e i Costantinidi. L'innovazione costantiniana, le sue radici e i suoi sviluppi.*

[3]   Hamilton, R.W. (1934) Excavations in the Atrium of the Church of the Nativity. *QDAP*, 3, 1–8.

[4]   Vincent, H. (1936–1937) Bethléem. Le sanctuaire de la Nativité d'après les fouilles récentes. *RBi*, 45, 544–574; 46, 93–121.

[5]   Bagatti, B. (1952) *Gli antichi edifici sacri di Betlemme in seguito agli scavi e restauri praticati dalla Custodia di Terra Santa (1948–1951)*. Jerusalem, Franciscan Printing Press.

[6]   Milani, G., Valente, M. & Alessandri, C. (2018) The narthex of the Church of the Nativity in Bethlehem: A non-linear finite element approach to predict the structural damage. *Computer & Structures*, ISNN 0045-7949, 207, 3–18.

[7]   NTC 2008 – Norme Tecniche per le Costruzioni – D.M. 14 Gennaio 2008.

[8]   Borri, A. & De Maria, A. (2009) L'indice di qualità muraria (IQM): evoluzione ed applicazione nell'ambito delle Norme Tecniche per le Costruzioni del 2008. *Proc. XIII Convegno Nazionale "L'Ingegneria Sismica in Italia"*, Bologna.

[9]   ARCO Software by P. Gelfi, University of Brescia, Italy ver.1.2 07/04/2008.

[10] Heyman, J. (1982) *The Masonry Arch*. Ellis Horwood, Chichester.

[11] HiStrA Structural Code ver. 3.0.1.

[12] PRO_SAP Professional Structural Analysis Program, 2SI, Software e Servizi per l'Ingegneria, Ferrara, Italy. Build 2014.07.0168 (ver. 13.0.0)-RY2014(b).

Chapter 10

Monitoring

C. Alessandri, N. Macchioni, M. Martinelli and B. Pizzo

Once the numerous and complex interventions of restoration and consolidation have been completed, a modern monitoring system can provide an effective tool for a real-time control of the state of conservation of the whole church and for a constant evaluation of the efficiency level of the interventions carried out, which is particularly important since non-traditional and innovative techniques of restoration and consolidation have been used. Moreover, a continuous monitoring allows to process and update the maintenance plan in real time and also to increase the level of knowledge of the global behavior of a complex structure as the one of the church.

The use of numerical techniques offers nowadays a valid contribute to predict with great accuracy the response of any kind of structure, whatever its degree of complexity, but the results that can be obtained cannot be considered reliable if not validated and improved by the data surveyed through a monitoring campaign able to record constantly the dynamic interaction between the structure and the environment (response to the wind, vibrations, earthquakes, changes of the subsoil conditions, displacements due to new load conditions etc.) and to highlight the occurrence of significant and unexpected events requiring a prompt intervention.

The monitoring of the structural response of the church, aimed at controlling the permanence over time of the structural performances both of the building as a whole and of the repairs and improvements made during the restoration, can benefit from the flow of data provided by the on-site monitoring system. Furthermore, the continuous flow of data and the possibility of evaluating them in almost real time allow, through a dedicated software, to define alert thresholds, on the basis of which the site manager can decide very quickly whether to keep the structure in service or not, as well as to schedule any safety measures.

The proposed monitoring system takes into account not only the interventions of repair and seismic improvement but also the studies carried out in previous survey campaigns and the acquired knowledge about the structural behavior of the church during the restoration works. It is a relatively complex system because it must control different parameters within the same historic building and respect, at the same time, some fundamental constraints such as sustainability, reduced invasiveness, reversibility and, obviously, reliability of the recorded data. It makes use of sensors like, for instance, accelerometers and velocimeters to evaluate acceleration and velocity, strain gauges and clinometers to define strains and slopes and thermal sensors and hygrometers to measure the temperature gradients and the amount of humidity in a defined environment.

The following are the monitoring systems provided for some basic components of the church.

10.1   Narthex

The monitoring system consists of (Figures 10.1, 10.2):

- wire strain gauges that measure the opening or closing of the vaults in correspondence of the so-called pulvini at the arch abutments,
- biaxial clinometers on the façade to measure possible rotations,
- accelerometers on the façade,
- strain gauge rosettes for monitoring diagonal tie rods of the horizontal steel frame,
- thermal sensors in the vault interspace,
- hygrometers in the vault interspace.

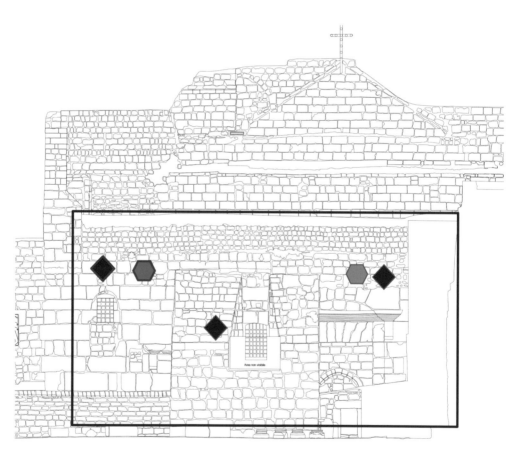

ACCELEROMETER

CLINOMETER

Figure 10.1 Monitoring system on the narthex façade.

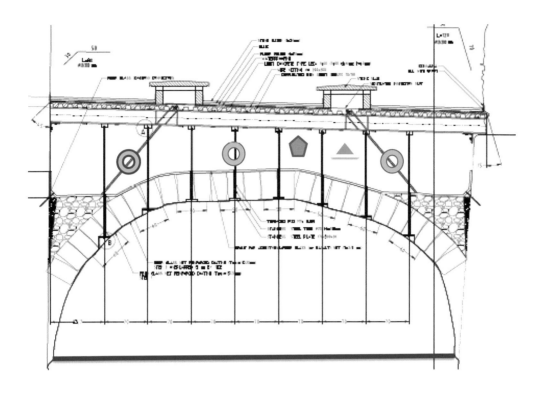

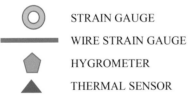

⬡ STRAIN GAUGE

━━━ WIRE STRAIN GAUGE

⬠ HYGROMETER

▲ THERMAL SENSOR

Figure 10.2 Internal monitoring system.

10.2 Church façade

The analyses and numerical simulations performed on the façade at each stage of the restoration design, including the definition of the seismic improvement interventions, have highlighted its vulnerability to the overturning in the presence of seismic actions. Therefore, the monitoring system may consist of (Figure 10.3):

* biaxial clinometers on the façade to measure rotations,
* accelerometers on the façade,
* weather station.

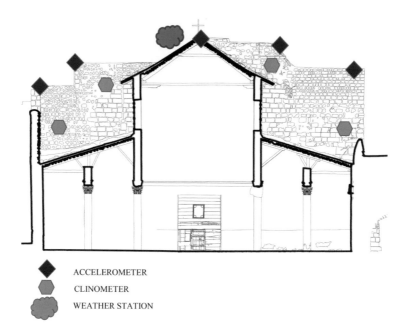

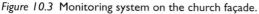

Figure 10.3 Monitoring system on the church façade.

10.3   Walls of the nave

The analyses and numerical simulations performed on these walls at each stage of the resto-ration design, including the definition of the seismic improvement interventions, have high-lighted their vulnerability to the overturning in the presence of seismic actions. Therefore, the monitoring system may consist of (Figure 10.4):

- biaxial clinometers to measure rotations,
- accelerometers.

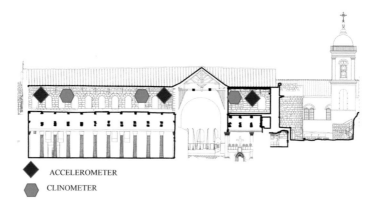

Figure 10.4 External monitoring system on the walls of the nave.

10.4   Orthodox wall (south wall)

It is the wall built at the Crusaders' time upon the external wall of the south aisles (Figures 10.5, 10.6). It is worth noting that this wall was raised simply in adherence with the orthogonal ones, although some decorative stones, still visible on both sides of the basilica and protruding from the transept wall, were probably used to improve the connection, which however remains rather weak. Moreover, the considerable length of this wall and the absence of intermediate retains make the wall highly vulnerable in the presence of a seismic event. In particular, the collapse of the wall might endanger the safety of visitors as well as damage the roof of the aisles, recently restored.

While waiting for an appropriate intervention of consolidation the following monitoring system can be provided (Figure 10.7):

*   biaxial clinometers to measure rotations,
*   accelerometers.

Figure 10.5 External view (south side).

Figure 10.6 Internal view (north side).

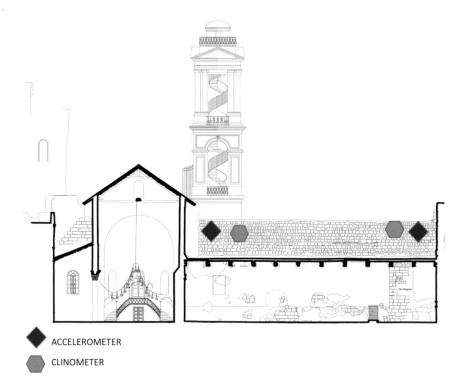

◆ ACCELEROMETER

⬡ CLINOMETER

Figure 10.7 Monitoring system on the internal (north) side of the wall.

10.5 Corners

For the problems mentioned in Chapter 8.5 the following monitoring system can be provided in both corners (Figure 10.8):

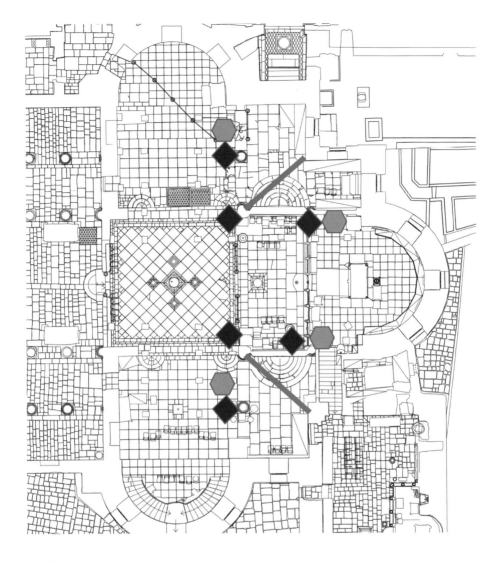

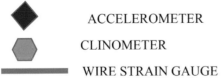

ACCELEROMETER

CLINOMETER

WIRE STRAIN GAUGE

Figure 10.8 Monitoring system in the corners.

- wire strain gauges to measure the opening or closing of the corners,
- biaxial clinometers to measure wall rotations,
- accelerometers on the walls,
- accelerometers on the transept corner columns.

10.6   Columns

The monitoring system may consist of (Figure 10.9):

- accelerometers on all the columns of both colonnades separating the nave from the aisles.

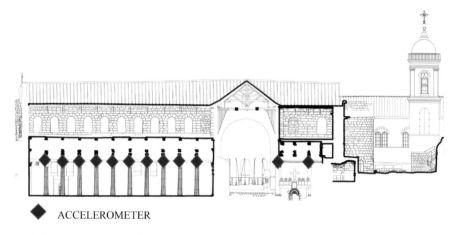

ACCELEROMETER

Figure 10.9 Monitoring system for the columns of the nave.

10.7   Wooden architraves

Surveys and checks carried out have shown that the architraves, which support the masonry walls of the nave, are subject to considerable compressive stresses perpendicular to the grain in the parts immediately above the capitals of the columns. The static checks, while being verified, have also shown that the present stress levels, even with permanent loads only, are in any case very high. There have also been more or less extensive phenomena of decay caused by biotic attacks or insects (although the interventions carried out have allowed to removing any cause of decay and to provide the most appropriate remedies). The provided monitoring system allows to control constantly the compressive stresses in correspondence of the support, stresses and/or strains in the middle of the span and also the wood moisture content.

All this information is useful to prevent and to keep under observation any fungal rot decay or even mechanical worsening of the wood. The monitoring system consists of (Figure 10.10):

- load cells in correspondence of the supports (columns),
- moisture content sensors in correspondence of the supports (columns),
- stress/strain sensors in the middle of each span.

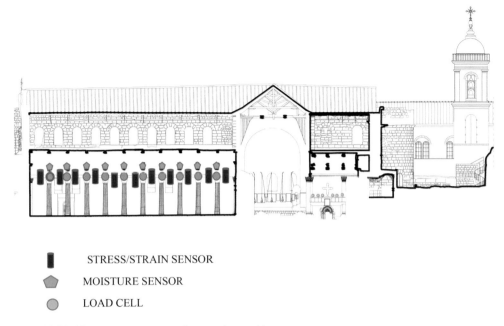

STRESS/STRAIN SENSOR

MOISTURE SENSOR

LOAD CELL

Figure 10.10 Monitoring system in the wooden architraves.

10.8   Roof wooden structures

Surveys and checks carried out have shown that the wooden structures of the roof (trusses, cantilevers, timber lacings, purlins, boards, joists etc.) of the church were affected by more or less extensive phenomena of degradation due to fungal rot or attacks by insects. The interventions carried out have allowed to adopt the most appropriate remedies and to restore the original safety conditions.

The proposed monitoring system allows keeping under constant control the stress level in the main structures (trusses) of nave, aisles, transept and central area, the behavior of glued or differently joined prostheses, the wood moisture contents, useful to prevent and keep under observation possible phenomena of wood degradation. It consists of (Figure 10.11):

- wood moisture content sensors at the support on the walls,
- strain/stress sensors on trusses and prostheses (in this latter case halfway between the original part of the beam and the prosthesis),
- clinometers and accelerometers on trusses (to detect as early as possible any contingent mechanical damage or disconnection).

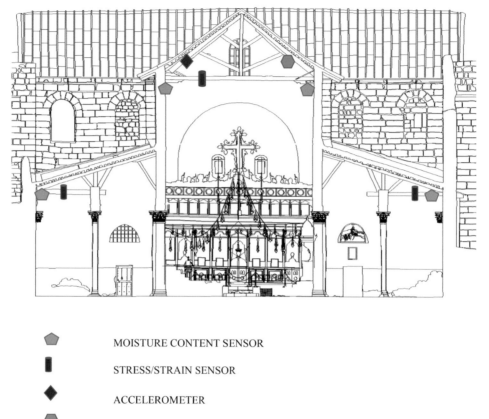

◯ MOISTURE CONTENT SENSOR

▮ STRESS/STRAIN SENSOR

◆ ACCELEROMETER

⬡ CLINOMETER

Figure 10.11 Monitoring system in the main wooden structures of the roof.

10.9   Fire and micro-climate detection system

A fixed automatic fire detection system has already been installed in order to detect and report a fire as quickly as possible (Fabio Altieri, Engineering & Consulting ABM Technology Group, Ferrara, Italy). The fire signal will be transmitted and displayed on a control and signal station.

An acoustic and visual alarm signal will be issued in all environments including the one affected by the fire. The purpose of this detection system is to:

- favor a timely displacement of people, and the evacuation, where possible, of goods;
- promptly activate the emergency evacuation intervention plans;
- activate the active protection systems against fire and any other safety measures.

All the components of the automatic fixed system, as required by UNI 9795, will comply with UNI EN 54–1. The system will include the following mandatory components:

- automatic fire detectors;
- manual reporting points;
- the control and signal station;
- power supply equipment;
- fire alarm devices.

Taking into account the presumably foreseen fire conditions and the type of combustible materials present in the spaces to be protected, linear smoke detectors or point detectors were installed, in combination with thermo-speed detectors.

The detection system is equipped with two power supply sources for electricity, primary and secondary, each of which can ensure the correct operation of the entire system, in compliance with UNI EN 54–4.

The fire detection system was designed according to the following technical standards:

UNI 9795: "Sistemi fissi automatici di rivelazione e di segnalazione allarme d'incendio";
UNI EN 54: "Fire Detection and Fire Alarm Systems";
EN 54–3: "Fire Detection and Fire Alarm Systems – Sounders";
EN 54–5: "Fire Detection and Fire Alarm Systems – Heat Detectors, Point Detectors";
EN 54–7: "Fire Detection and Fire Alarm Systems – Smoke Detectors, Point Detectors";
EN 54–1: "Fire Detection and Fire Alarm Systems – Manual call point";
EN 54–12: "Fire Detection and Fire Alarm Systems – Smoke Detectors – Line detectors using an optical light beam";
EN 54–17: "Fire Detection and Fire Alarm Systems – Short Circuit Isolators";
EMC Directive 2004/108/EC;
REACH – Regulation (EC) No 1907/2006;
ROHS – Directive 2011/65/EU;
WEEE – Directive 2012/19/EU

The microclimatic detection system (Eng. Elena Mainardi, Ferrara, Italy) was designed according to the UNI 10829/1999, UNI EN 15758/2004 and UNI EN 16242/2013 Standards, which were taken as general references for choosing the number and positioning of the microclimatic sensors. They will constantly detect thermo-hygrometric and lighting parameters in order to monitor and highlight any microclimatic problems that might cause deterioration of the artefacts present inside the building.

The data will be taken from the sensors and stored in a unit connected to a computer, as well as transferred even outside, to another internet-connected station. The choices related to the number of sensors aim at focusing attention on the areas of the basilica with the presence of artefacts of high interest or with a high concentration of tourists. It is worth noting that the microclimatic detection is not linked, in this specific case, to the creation of a microclimate control system but only to a data evaluation and that the positioning of the sensors will strongly depend on the location of the artefacts themselves and on the aesthetic choices.

The following Technical Standards apply to this microclimatic detection system:

UNI 10829:1999: "Beni di interesse storico artistico: condizioni ambientali di conservazione. Misurazione e analisi"
EN 15758:2010 "Conservation of cultural property. Procedures and instruments for measuring temperatures of the air and the surfaces of objects"

EN 16242:2013 "Conservation of cultural heritage. Procedures and instruments for measuring humidity in the air and moisture exchanges between air and cultural property"

Both fire detection and microclimatic systems will be maintained in the conditions of full
efficiency by providing for constant data minitoring, the maintenance of the systems and all
the necessary periodic inspections.

In addition, it is necessary to keep a special register updated, available to the competent
authorities, with the signature of the managers and with the following annotations:

- work carried out on systems or in supervised areas, such as renovations, structural modifications etc., if these can influence the efficiency of the systems themselves;
- tests performed;
- faults suffered by the systems and their causes, as well as the procedures activated to
avoid their recurrence;
- interventions in the event of fire: the number of detectors that came into operation, the
manual signal points used, the causes of the fire itself and any other information useful
for evaluating the efficiency of the entire plant will be recorded.

All the systems will be subject to inspection and maintenance at least twice a year with an
interval between the two of not less than 5 months. These operations will be carried out
only by expert and qualified personnel and will be regularly formalized in the aforemen-
tioned register, highlighting any shortcomings or anomalies found with respect to the last
verification.

Index